R. L. PLACKETT
PROFESSOR OF STATISTICS
UNIVERSITY OF NEWCASTLE UPON TYNE

AN INTRODUCTION TO THE THEORY OF STATISTICS

OLIVER & BOYD EDINBURGH

OLIVER AND BOYD
Tweeddale Court
14 High Street
Edinburgh EH1 1YL

A Division of Longman Group LIMITED

UMT 40
ISBN 0 05 002332 2
First published 1971

Printed in Great Britain by
R. & R. Clark Ltd., Edinburgh

UNIVERSITY MATHEMATICAL TEXTS

EDITORS/ALAN JEFFREY AND IAIN T. ADAMSON

AN INTRODUCTION TO THE THEORY OF STATISTICS

RECENT UNIVERSITY MATHEMATICAL TEXTS

An Introduction to Modern Mathematics/
Albert Monjallon
Quantum Mechanics/R. A. Newing and J. Cunningham
Probability/J. R. Gray
Series Expansions for Mathematical Physicists/H. Meschkowski
Calculus of Variations/J. C. Clegg
Elementary Rings and Modules/Iain T. Adamson

PREFACE

One of the earliest volumes in this series is' Statistical mathematics' by the late Professor A. C. Aitken. Like everything he wrote, it is a model of conciseness and elegance. During the thirty years which has elapsed since his book was first published, the scope of statistics has been enormously expanded, and a different view is now possible of what constitutes the essential features of the subject. I owe to Professor A. Jeffrey the suggestion that the University Mathematical Texts should include a volume written from this viewpoint.

Statistics is concerned with the collection and analysis of data. Statistical methods always require common sense and clarity of thought. They often need, in addition, the theoretical basis which only mathematics can supply. The object of this book is to provide a first course in the mathematical theory of statistics for students of mathematics at institutions of higher education. Readers should have a thorough understanding of differential and integral calculus, together with a grasp of the elements of vector spaces. A minimal knowledge of convergence and complex variables is also useful.

There is a core of material which selects itself, consisting of what is presented in the first seven chapters. The remainder of the book forms a personal choice, and the approach which has been followed is to motivate and explain some of the statistical methods which are most useful in practice. Many important topics have been excluded for reasons of space, for example, experimental design and the analysis of time series, as well as further generalizations of the problems discussed and other approaches to their solution. At the chosen level of presentation, complete proofs are possible for nearly all the results in the first nine chapters. The remaining two are concerned with more difficult topics, but are justified by the importance of the material in the final chapter. Here it has sometimes been necessary to sketch proofs or omit them altogether. The exercises at the end of each chapter are constructed largely as an extension of the main text, and readers are urged to work through them. A course of practical work is also necessary before the methods described can be fully appreciated or applied with any confidence.

The excellent typing of Miss J. Edger and Mrs. M. Smith has been invaluable in preparing the material for publication. I am solely responsible for all the defects of the book and would be glad to have notification of them.

CONTENTS

CHAPTER 1
STATISTICS AND PROBABILITY

§1. Introduction

Nearly every phase of human activity has its statistical aspects and numerical information of all kinds is made available in ever increasing quantities. Newspapers regularly list figures relating to road accidents, numbers of votes cast in elections, financial details about industry and commerce, and results from football, cricket or other sports. The publications of governments, and governmental bodies like the United Nations, contain a very extensive range of statistical data, largely in the fields of economic and social studies. A consequence of the emphasis given to the purely factual side of statistics is the popular belief that the subject is synonymous with groups of numerical data. The time when this belief was correct has long since gone, although the numerical aspect remains important. Statistics is now an organized discipline, the object of which is to provide proper methods for the collection and analysis of data, whatever is the field of enquiry.

The fact that special methods are necessary is indicated by the following examples, which have been selected to represent a wide range of possibilities.

Medicine. A hospital has a ward containing emergency beds and the number which are occupied varies from day to day. The problem is to give a numerical summary description of the variation in occupancy, and to compare the description for different medical specialties.

Geology. Measurements are made of the concentration of copper in parts per million found in ores where an extraction plant may be established. A considerable variability is observed, and since concentration is measured on a continuous scale, the variation is different in character from the previous example, where the number of occupied beds is necessarily an integer. We require a summary description in numerical terms of the way in which concentration varies.

Sociology. Prior to a parliamentary election with candidates from three parties, an attempt is made to forecast the final percentages of the poll by asking a sample of 2000 people which way they will vote. The sociologist has to decide how the sample of 2000 should be taken so as to be representative of the 55,000 people in the constituency, and he needs to know how to calculate the likely error in the forecast.

Physics. A propane bubble chamber is exposed to a beam of mesons and observations are made of the angle θ which the decay pions make with a fixed direction. The values of $\cos\theta$ appear to be scattered fairly uniformly over the interval $(-1, 1)$. The research worker requires to give precise expression to the hypothesis of uniformity and examine how far the data agree with it.

Traffic studies. Records of the number of fatal accidents in a ten-mile stretch of main road can be subdivided depending on whether they refer to the road at the beginning of the survey, or after a speed restriction was imposed, or again after widening and the construction of a central reservation. The county council responsible wish to see how far, if at all, these improvements have affected the accident rate, when allowance is made for the differing periods of use of the road in its various states, the general increase in traffic flow, and seasonal variation.

Agriculture. Barley is grown on experimental plots of equal area, which are divided into groups of equal size. A different quantity of a nitrogenous fertilizer is applied to each group, one group receiving none at all. When the yields of grain become available, they vary between plots which have been given the same treatment as well as between treatments. The question is to describe how the effect of the fertilizer varies with the quantity used, and then make recommendations for applying it in practice.

Technology. An industrial process produces large numbers of screws which are usually identical but may occasionally be defective as a result of errors in machine setting or tool wear. Screws are collected into boxes before despatch to the purchaser, and a few are selected at random from each box in order to assess the proportion of defective screws. As a result, each box is either accepted as satisfactory or rejected as unsatisfactory. The manufacturer has to decide how many screws to inspect, taking into account the cost of inspection and the need to avoid, as far as possible, both the rejection of satisfactory boxes and the acceptance of unsatisfactory boxes.

Hydrology. A large industrial complex requires more water than the local towns can supply and plans are made to construct a reservoir in a nearby hilly region. Rainfall records for the valley selected are available for the past 30 years and it can be supposed that much of this water will be retained when the dam is built. The water engineer has to decide how large the reservoir should be in order to accommodate the cumulative run-off for a period of years and allow for considerable fluctuations in rainfall, but nevertheless supply a specified and increasing demand.

Economics. A sequence of measurements records the weekly amount of electricity consumed in a large town. The general features of the data consist of a steady increase in the long term, a fairly regular oscillation as consumption rises in the winter and falls in the summer, and irregular oscillations for which no obvious reason is known. We require to characterize the main features in quantitative terms, so as to predict future values of consumption and estimate the length of time for which the present generating facilities will be adequate.

In every one of these illustrations, it is an observational fact that repetitions of the experiment under what are apparently identical conditions do not give rise to the same measurements, and the variability has to be accepted because nothing can be done to remove it. However, the measurements often appear consistent with differences of a systematic nature, as for instance a reduction in accident rate following road improvement, or an increase in yield of grain from the use of fertilizer. The general problem is to separate out such systematic effects from the variability in which they are embedded, and the problem can only be solved by introducing new ideas and procedures which are different in principle from those used in ordinary reasoning. The special methods required will be introduced in two stages. First, a theory of probability is motivated by experiments of the kind described above and the consequences of an axiomatic formulation are explored. Secondly, techniques of statistical analysis are developed which illustrate how the information provided by variable measurements can be assessed.

§2. Frequency and probability

Suppose that the students who first enter a University in a given year are listed in alphabetical order, and that the lists for every year are joined in chronological order. Each student is associated with various characteristics such as month of birth, number of brothers and sisters, or height on entry, all of which can, in principle, be recorded against his or her name. For any such characteristic we thus obtain a sequence of observations which are effectively those obtained from a repeated experiment in which the student is the experimental unit.

Consider the month of birth, and let a_n be the number of students born in January among the first n students. Thus $a_1 = 1$ or 0 depending on whether or not the first student was born in January. Denote the proportion of January births among the first n students by r_n, so that

$$r_n = a_n/n.$$

More briefly, a_n is known as the **frequency** of January births and r_n as the **frequency-ratio**. Since a_n must lie between 0 and n, we obtain

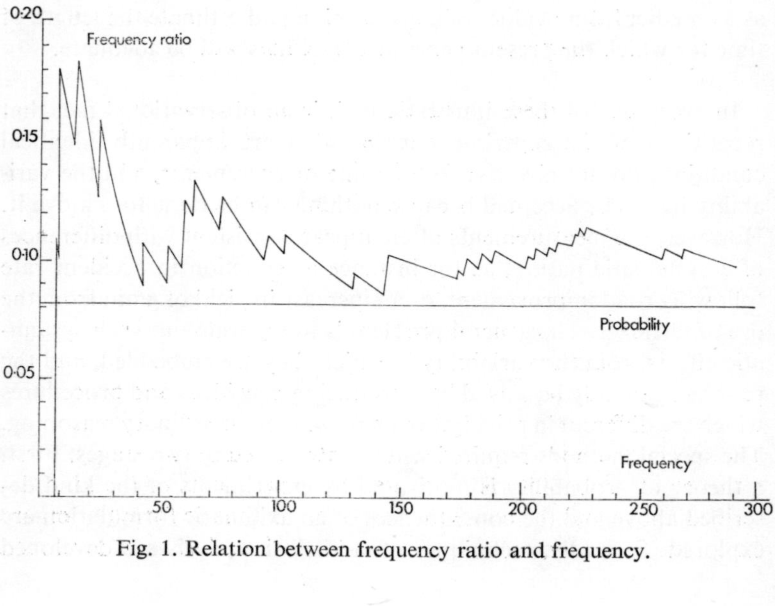

Fig. 1. Relation between frequency ratio and frequency.

$$0 \leqslant r_n \leqslant 1.$$

Construct a graph of r_n as a function of n. It would almost certainly be found that this graph has an appearance similar to that shown in Fig. 1. Initially $r_1 = 1$ or 0, and for small or moderate values of n there are considerable fluctuations, but as n increases the amplitude of the fluctuations diminishes and eventually r_n appears to be tending to a limit. For any other University, the resulting graph of r_n as a function of n will have the same general appearance, and the frequency-ratios $\{r_n\}$ will approach the same limit, although any two such graphs can be distinguished by detailed inspection. The value at which the frequency-ratios eventually stabilize is known as the **probability** of a January birth.

Similar considerations apply in respect of any other month. For the first n students, let the frequencies of birth in different months and the corresponding frequency-ratios be as follows.

Month	Jan.	Feb.	March	April	...	Dec.
Frequency	a_n	b_n	c_n	d_n	...	f_n
Frequency-ratio	r_n	s_n	t_n	u_n	...	w_n

A graph of s_n as a function of n is once again similar in appearance to Fig. 1, with fluctuations which eventually diminish as a limiting value is approached. The limit here is the probability of a February birth, and the corresponding graphs for $t_n, u_n, \ldots, w_n$ approach limits which are the probabilities of birth in March, April, . . . , December.

Denote the probabilities of birth in January, February, . . . December by $p_1, p_2, \ldots, p_{12}$ respectively. Each of the frequency-ratios $r_n, s_n, \ldots, w_n$ lies in [0, 1] for all n, and the limits of the corresponding sequences must have the same property, so that

$$0 \leqslant p_i \leqslant 1 \quad (i = 1, 2, \ldots, 12).$$

Each student is certainly born in one of the twelve months, and therefore

$$a_n + b_n + \ldots + f_n = n,$$

or equivalently

$$r_n + s_n + \ldots + w_n = 1.$$

Proceeding to the limit, we infer that

$$p_1 + p_2 + \ldots + p_{12} = 1.$$

The time of year when a student is born can also be classified by the quarter into which it falls. Denote the frequency of births in January, February or March among the first n students by g_n and the frequency-ratio by x_n. Then

$$g_n = a_n+b_n+c_n$$

or equivalently

$$x_n = r_n+s_n+t_n.$$

The probability of birth in the first quarter, say q_1, is thus

$$q_1 = p_1+p_2+p_3,$$

and the probabilities for the other quarters are similarly obtained. Such results are more readily appreciated if written using the notation

$$p(\text{Birth in the first quarter}) \\ = p(\text{Birth in Jan.})+p(\text{Birth in Feb.})+p(\text{Birth in Mar.}),$$

where $p(.)$ is read as the probability of the statement contained within the brackets.

Consider next the subsequence formed from the original sequence by selecting only those students who were born in the first quarter of the year. Among the first g_n such students there are a_n who were born in January. A graph of a_n/g_n as a function of n exhibits the same general features as Fig. 1. The limit approached by a_n/g_n is described as the probability of a January birth, given that the birth is in the first quarter, and it is written as

$$p(\text{Birth in January} \mid \text{Birth in first quarter}).$$

The qualification that the birth is in the first quarter is a condition under which the limit was obtained, and such limits are described as **conditional probabilities.** Now

$$a_n/g_n = (a_n/n)/(g_n/n),$$

and the limit of the ratio on the right is p_1/q_1. Hence

$$p(\text{Birth in January} \mid \text{Birth in first quarter}) \\ = p(\text{Birth in January})/p(\text{Birth in first quarter}).$$

For a reason which emerges shortly, this expression of a conditional probability is rather specialized, and the concept can be further elucidated by another example.

Each student can be classified by sex, and by whether or not height exceeds 5 ft. 6 in. The four possible outcomes and the frequencies with which they appear among the first n students are as follows.

Height greater than 5 ft. 6 in. and Male:	h_n
Height 5 ft. 6 in. or less and Male:	i_n
Height greater than 5 ft. 6 in. and Female:	j_n
Height 5 ft. 6 in. or less and Female:	k_n.

Thus the frequency of males is h_n+i_n. Take the limit on each side of the identity

$$h_n/(h_n+i_n) = (h_n/n)/\{(h_n+i_n)/n\},$$

and we obtain

$$p(\text{Height greater than 5 ft. 6 in.} \mid \text{Male}) = p(\text{Height greater than 5 ft. 6 in. and Male})/p(\text{Male}).$$

An expression of this type holds generally for conditional probabilities. The reason why the previous conditional probability was simpler in form is because a birth in January implies a birth in the first quarter, whence

$$p(\text{Birth in January and Birth in first quarter}) = p(\text{Birth in January}).$$

§3. Axioms of probability

We cannot prove that frequency-ratios behave in the way illustrated by Fig. 1, although there is much evidence that they generally do so. The limiting value of a sequence of frequency-ratios $\{r_n\}$, say p, cannot be defined in a mathematical sense. This would require, for any $\varepsilon > 0$, that $|r_n-p| < \varepsilon$ for all n greater than a value depending on ε, whereas a sequence of results which break the inequality is always possible. Examples of the type that we have considered, while they supply an invaluable background to the theory of probability, cannot provide it with a rigorous foundation. We therefore express the basic properties of probability in the form of axioms, which are suggested by the results of repeated experiments but are independent of whether or not our inferences from these experiments are correct.

The possible experimental results can always be classified by an enumerable set of **outcomes** $\omega_0, \omega_1, \omega_2, \ldots$ which are such that (*a*) no two outcomes ω_i and ω_j can be observed simultaneously, and (*b*) the outcomes taken together account for all the possibilities. Various classifications are usually available, and the choice between them depends on the degree of refinement that is required.

Example 1. Time of birth in a specified leap year can be classified by four quarters, twelve months, or 366 days.

Example 2. Concentration of copper certainly lies between 0 and 1 million ppm, and can be classified by contiguous equal intervals of 0·1 or 1 or 5 . . . ppm extending over the full range.

Example 3. Sex and height taken together can be classified by the four outcomes listed in the discussion of conditional probability. It would be more usual for height to be classified by equal intervals on the scale of measurement, in which case the number of outcomes would be much larger than four.

An **event** is defined as a set of outcomes. Let A be the event which consists of the outcomes $\omega_i, \omega_j, \ldots, \omega_k$. We say that A **occurs** if any of $\omega_i, \omega_j, \ldots, \omega_k$ are observed. The **complement** of A is denoted by $\bar{A}$ and consists of all outcomes not in A. Thus $\bar{A}$ occurs if A does not occur. The set which contains no outcomes is the **null event**, denoted by $\varnothing$, and the set of all outcomes is the **outcome space**, denoted by Ω.

Let A_1 and A_2 be any two events. The notation and terminology of set theory can be used to form combinations of A_1 and A_2. Thus the event which is comprised of all outcomes contained in either A_1 or A_2 corresponds to the union of the sets A_1 and A_2, and is denoted by $A_1 \cup A_2$. We say that $A_1 \cup A_2$ occurs if at least one of A_1 and A_2 occurs. Similarly the event which is comprised of all outcomes contained in both A_1 and A_2 corresponds to the intersection of the sets A_1 and A_2, and is denoted by $A_1 \cap A_2$. We say that $A_1 \cap A_2$ occurs if both A_1 and A_2 occur. The events A_1 and A_2 are said to be **exclusive** if $A_1 \cap A_2 = \varnothing$.

Example 4. A_1 is the event that birth lies in the first four months of the year, and A_2 that it lies in the second quarter. Then $A_1 \cup A_2$ refers to births in the first six months, and $A_1 \cap A_2$ to births in April. The outcomes here can be birth in January, birth in February, . . . , or births in any unit of time smaller than one month.

Example 5. A_1 is the event that copper concentration lies between 20 and 40 ppm, and A_2 that it lies between 30 and 60 ppm. Then $A_1 \cup A_2$ refers to all concentrations between 20 and 60 ppm, and $A_1 \cap A_2$ to all between 30 and 40 ppm.

Example 6. A_1 is the event: male with height between 5 ft. 3 in. and 5 ft. 9 in.; and A_2 is the event: height between 5 ft. 6 in. and 6 ft. Then $A_1 \cup A_2$ is the event: male with height between 5 ft. 3 in. and

6 ft., or female with height between 5 ft. 6 in. and 6 ft.; and $A_1 \cap A_2$ is the event: male with height between 5 ft. 6 in. and 5 ft. 9 in.

These ideas extend naturally to any finite number of events. Thus the union of the sets $A_1, A_2, \ldots$ is denoted by $\cup A_i$, and we say that the event $\cup A_i$ occurs if at least one of the events $A_1, A_2, \ldots$ occurs. Similarly the intersection of the sets $A_1, A_2, \ldots$ is denoted by $\cap A_i$, and we say that the event $\cap A_i$ occurs if all of the events $A_1, A_2, \ldots$ occur. If $A_i \cap A_j = \varnothing$ whenever $i \neq j$, then the events $\{A_i\}$ are described as **mutually exclusive**. When $\cup A_i = \Omega$, the events $\{A_i\}$ are said to be **exhaustive**, since one of them must necessarily occur.

With every event A is associated a number $p(A)$, called the **probability** of A and such that the following conditions are satisfied.

Axiom 1. *The probabilities of the outcomes* $\{\omega_j\}$ *are non-negative real numbers such that*

$$\Sigma p(\omega_j) = 1.$$

This condition is known as the **law of total probability.**

Axiom 2. *Let A be the event which consists of the outcomes* $\omega_i, \omega_j, \ldots, \omega_k$. *Then*

$$p(A) = p(\omega_i) + p(\omega_j) + \ldots + p(\omega_k).$$

Axiom 3. *Given two events A and B, with* $p(A) \neq 0$, *the* **conditional probability** *of B given A is denoted by* $p(B \mid A)$ *and defined by*

$$p(B \mid A) = p(A \cap B)/p(A).$$

§4. Consequences of the axioms

An immediate deduction from axioms 1 and 2 is that $p(\Omega) = 1$. The null event contains no outcomes, whence $p(\varnothing) = 0$. Thus the probability of A satisfies

$$0 \leqslant p(A) \leqslant 1.$$

Let $A_1, A_2, \ldots$ be mutually exclusive events. Each outcome in $\cup A_i$ is contained in exactly one of the events $\{A_i\}$. We therefore obtain from axiom 2 that

$$p(\cup A_i) = \Sigma p(A_i).$$

This is known as the **addition law of probabilities for mutually exclusive events.** If the $\{A_i\}$ are not only mutually exclusive but also exhaustive, then $\cup A_i = \Omega$, whence $\Sigma p(A_i) = 1$. In particular,

$$p(A) + p(\bar{A}) = 1.$$

Suppose that the events $\{B_i\}$ are mutually exclusive and exhaustive. Let A be an arbitrary event. The events $A \cap B_i$ and $A \cap B_j$ are mutually exclusive when $i \neq j$, because otherwise B_i and B_j would have outcomes in common. Since $\cup B_i = \Omega$, each outcome in A is contained in one of the $\{B_i\}$, and hence $A = \cup(A \cap B_i)$. Using the addition law for mutually exclusive events, we obtain

$$p(A) = \Sigma p(A \cap B_i).$$

The proof of a result such as this can be elucidated by a diagram where events are represented by regions in a plane and their probabilities by the corresponding areas. In Fig. 2, $B_1, B_2, \ldots$ are represented by rectangles, and A by an elliptical shape.

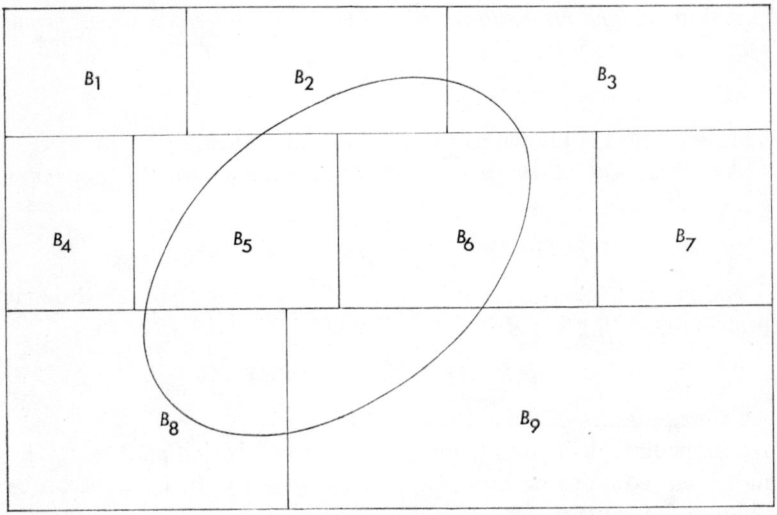

Fig. 2. Diagram for events and probabilities.

Example 1. Students are classified by sex, and by whether height is less than 5 ft. 6 in., between 5 ft. 6 in. and 6 ft., or above 6 ft. The probabilities for the six resulting outcomes are given below.

	Less than 5 ft. 6 in.	*Between 5 ft. 6 in. and 6 ft.*	*Above 6 ft.*
Male	0·18	0·27	0·15
Female	0·14	0·20	0·06

Let A be the event that a student is male and denote the three categories of height by B_1, B_2, B_3 respectively. The probabilities in the first row of the table are $p(A \cap B_1)$, $p(A \cap B_2)$, and $p(A \cap B_3)$ respectively, whence

$$p(A) = 0{\cdot}18+0{\cdot}27+0{\cdot}15 = 0{\cdot}60.$$

The definition of conditional probability given by axiom 3 can be written in the form

$$p(A \cap B) = p(A)p(B \mid A).$$

When the factorization $p(A \cap B_i) = p(B_i)p(A \mid B_i)$ is applied to each term of $\Sigma p(A \cap B_i)$, we obtain

$$p(A) = \Sigma p(B_i)p(A \mid B_i).$$

The alternative factorization is $p(A \cap B_i) = p(A)p(B_i \mid A)$, whence

$$p(B_i \mid A) = p(B_i)p(A \mid B_i)/p(A).$$

Substituting for $p(A)$, we find that

$$p(B_i \mid A) = p(B_i)p(A \mid B_i)/\Sigma p(B_i)p(A \mid B_i).$$

This result is known as **Bayes's theorem.**

Example 2. Male students are classified by whether height is less than 5 ft. 6 in., between 5 ft. 6 in. and 6 ft., or above 6 ft. The probabilities for the three outcomes are respectively 0·30, 0·45, and 0·25. Female students are similarly classified and the corresponding probabilities are 0·35, 0·50, and 0·15 respectively. Let B_1 refer to male students, B_2 to female, and A to heights above 6 ft. Suppose that $p(B_1) = 0{\cdot}60$, whence $p(B_2) = 0{\cdot}40$ by total probability. Then

$$p(A) = (0{\cdot}60 \times 0{\cdot}25)+(0{\cdot}40 \times 0{\cdot}15) = 0{\cdot}21.$$

Given that a student is male, the conditional probability that height exceeds 6 ft. is 0·25, as compared with the unconditional probability of 0·21 for heights above 6 ft. The difference between the two probabilities simply reflects the fact that male students tend to be taller than females. Given that the height of a student exceeds 6 ft., the conditional probability that a student is male can be obtained from Bayes's theorem and is

$$(0{\cdot}60 \times 0{\cdot}25)/0{\cdot}21 = 0{\cdot}71 \text{ approximately.}$$

The numbers chosen here are all consistent with those of example 1.

Successive applications of axiom 3 show that

$$p(A \cap B \cap C) = p(A \cap B)\, p(C \mid A \cap B)$$
$$= p(A)\, p(B \mid A)\, p(C \mid A \cap B).$$

The generalization to events $A_1, A_2, \ldots, A_n$ takes the form

$$p(A_1 \cap A_2 \cap \ldots \cap A_n)$$
$$= p(A_1)\, p(A_2 \mid A_1)\, p(A_3 \mid A_1 \cap A_2) \ldots$$
$$p(A_n \mid A_1 \cap A_2 \cap \ldots \cap A_{n-1}).$$

We proceed to describe conditions under which the formula for the joint occurrence of several events is greatly simplified.

§5. Independence

This fundamental concept underlies the greater part of statistical theory. Suppose that the students entering a university are classified according to height, and that the outcomes include one for heights below 5 ft. Students can also be classified according to the distance between home and university, in such a way that there is an outcome for distances under 50 miles and another for distances over 100 miles. List the students in alphabetical order by years as in previous uses of this example, and consider the subsequence formed by students who are less than 50 miles from home. There is clearly a connection between height and sex, as noted in the previous section, but height and distance appear to be completely unrelated. In the absence of an association between height and distance, the limit of the frequency-ratio for heights below 5 ft. is the same in the subsequence where distances are less than 50 miles as in the full sequence. We thus obtain

$$p(\text{Height below 5 ft.} \mid \text{Distance below 50 miles})$$
$$= p(\text{Height below 5 ft.}).$$

By the definition of conditional probability, the same result can be expressed in symmetrical form as

$$p(\text{Height below 5 ft. and Distance below 50 miles})$$
$$= p(\text{Height below 5 ft.})\, p(\text{Distance below 50 miles}).$$

Similar results hold for other outcomes, as for example that

$$p(\text{Height below 5 ft. and Distance above 100 miles})$$
$$= p(\text{Height below 5 ft.})\, p(\text{Distance above 100 miles}).$$

We now proceed to formal definitions. A **method of classification** is defined by any system of mutually exclusive and exhaustive events. Suppose that events $\{A_i\}$ define one method of classification, and that events $\{B_j\}$ define another. A third method of classification is then defined by the events $\{A_i \cap B_j\}$. We say that the methods of classification $\{A_i\}$ and $\{B_j\}$ are **independent** if

$$p(A_i \cap B_j) = p(A_i)\, p(B_j) \text{ for all } i \text{ and } j.$$

A natural extension to several methods of classification, defined by $\{A_i\}$, $\{B_j\}$, . . . , $\{C_k\}$, is expressed by the requirement that

$$p(A_i \cap B_j \cap \ldots \cap C_k) = p(A_i)p(B_j) \ldots p(C_k) \text{ for all } i, j, \ldots, k,$$

in which case we say that the classifications are **mutually independent**. However, two difficulties arise. First, there are few practical examples of mutually independent classifications. Secondly, in the absence of mutual independence, the analysis of a multiple classification presents complex problems. We therefore turn to another class of experimental situations where the concept of independence is readily acceptable and analytically fruitful.

Suppose that a list of students in alphabetical order by years is available at each of two universities. For convenience, we specifically refer to the Universities of Dee and Tay. Let the lists be arranged in parallel, with the nth Dee student next to the nth Tay student for all n. In the Dee list, the frequency-ratio for students below 5 ft. in height tends to the limit p_0, and that for students above 6 ft. to the limit p_1. We naturally expect, and will assume, that the corresponding limits in the Tay list are also p_0 and p_1.

Now pick out all the students in the Tay list who are next to a Dee student with height below 5 ft., and consider the properties of the subsequence thus formed, in particular the limit for the frequency-ratio of Tay students with heights above 6 ft. Since there is no conceivable connection between the two lists, we feel intuitively that this limit would be p_1, the same value as in the full sequence of Tay students. Experimental evidence would almost certainly support our assumption. Thus

$$\begin{aligned} &p(\text{Tay student above 6 ft.} \mid \text{Dee student below 5 ft.}) \\ &\qquad = p(\text{Tay student above 6 ft.}). \end{aligned}$$

Writing the result in symmetrical form as before, we have

$$\begin{aligned} &p(\text{Tay student above 6 ft. and Dee student below 5 ft.}) \\ &\qquad = p(\text{Tay student above 6 ft.})\, p(\text{Dee student below 5 ft.}). \end{aligned}$$

Similar results hold for the other outcomes.

In general terms, let the outcomes of an experiment be $\omega_0, \omega_1, \omega_2, \ldots$ with probabilities $p(\omega_0), p(\omega_1), p(\omega_2), \ldots$ respectively. When the experiment is performed twice, the possible outcomes consist of all pairs of the form (ω_i, ω_j), and their probabilities are denoted by $p(\omega_i, \omega_j)$. We say that the two experiments are **independent** when

$$p(\omega_i, \omega_j) = p(\omega_i)p(\omega_j) \text{ for all } i \text{ and } j.$$

The system of probabilities $\{p(\omega_i)p(\omega_j)\}$ is consistent with axiom 1 because $p(\omega_i)p(\omega_j)$ is a non-negative real number and

$$\Sigma\Sigma p(\omega_i)p(\omega_j) = \Sigma p(\omega_i)\Sigma p(\omega_j) = 1.$$

Let A_1 be any event in the first experiment and A_2 any event in the second. Denote the outcomes in A_1 by $\{a_{1i}\}$ and those in A_2 by $\{a_{2j}\}$. The event consisting of A_1 in the first experiment followed by A_2 in the second is denoted by (A_1, A_2) and is comprised of all the outcomes (a_{1i}, a_{2j}). According to axiom 2,

$$p(A_1, A_2) = \Sigma\Sigma p(a_{1i}, a_{2j}).$$

When independence obtains

$$\Sigma\Sigma p(a_{1i}, a_{2j}) = \Sigma\Sigma p(a_{1i})p(a_{2j}) = \Sigma p(a_{1i})\Sigma p(a_{2j}),$$

and we conclude that

$$p(A_1, A_2) = p(A_1)p(A_2).$$

These ideas readily extend in the following way. Consider a sequence consisting of n experiments which may be all different. Denote the outcomes of the first experiment by $\{\omega_{1i}\}$, those of the second by $\{\omega_{2j}\}, \ldots$, and those of the nth by $\{\omega_{nk}\}$. The possible outcomes for the sequence are all combinations of the form $(\omega_{1i}, \omega_{2j}, \ldots, \omega_{nk})$. A sequence of n **mutually independent experiments** is defined by the condition that

$$p(\omega_{1i}, \omega_{2j}, \ldots, \omega_{nk}) = p(\omega_{1i})p(\omega_{2j}) \ldots p(\omega_{nk}) \text{ for all } i, j, \ldots, k.$$

Given that A_1 occurs in the first experiment, A_2 in the second, $\ldots$, and A_n in the nth, we have

$$p(A_1, A_2, \ldots, A_n) = p(A_1)p(A_2) \ldots p(A_n).$$

This is known as the **multiplication law of probabilities for the events of mutually independent experiments.**

EXERCISES 1

1. Show that

$$p(A) = p(A \cap B)+p(A \cap \bar{B}) \text{ and } p(A \cup B) = p(B)+p(A \cap \bar{B}).$$

Deduce that

$$p(A \cup B) = p(A)+p(B)-p(A \cap B).$$

2. Given two events A and B, prove the following results.

$$p(\text{Neither occurs}) = 1-p(A \cup B).$$
$$p(\text{Exactly one occurs}) = p(A \cup B)-p(A \cap B).$$
$$p(\text{At most one occurs}) = 1-p(A \cap B).$$

3. Show that the methods of classification defined by A, $\bar{A}$ and B, $\bar{B}$ are independent if

$$p(A \cap B) = p(A)p(B).$$

Prove that an equivalent condition is

$$p(A \cap B)p(\bar{A} \cap \bar{B}) = p(A \cap \bar{B})p(\bar{A} \cap B).$$

4. Show that

$$p\{A \cap (B \cup C)\} = p(A \cap B)+p(A \cap C)-p(A \cap B \cap C).$$

From the result of exercise 1 applied to $p\{A \cup (B \cup C)\}$, deduce that

$$p(A \cup B \cup C) = p(A)+p(B)+p(C)$$
$$-p(A \cap B)-p(B \cap C)-p(C \cap A)+p(A \cap B \cap C).$$

5. Given three events A, B, and C, prove the following results.

$p(\text{At most two occur})$
$$= 1-p(A \cap B \cap C).$$

$p(\text{At least two occur})$
$$= p(A \cap B)+p(B \cap C)+p(C \cap A)-2p(A \cap B \cap C).$$

$p(\text{Exactly two occur})$
$$= p(A \cap B)+p(B \cap C)+p(C \cap A)-3p(A \cap B \cap C).$$

6. Show that the probabilities of the events in the previous exercise, conditional on the occurrence of A, are as follows.

$$p(\text{At most two} \mid A) = 1 - p(A \cap B \cap C)/p(A).$$

$$p(\text{At least two} \mid A) = \{p(A \cap B) + p(A \cap C) - p(A \cap B \cap C)\}/p(A).$$

$$p(\text{Exactly two} \mid A) = \{p(A \cap B) + p(A \cap C) - 2p(A \cap B \cap C)\}/p(A).$$

7. We say that the methods of classification defined by A, $\bar{A}$ and B, $\bar{B}$ are conditionally independent given C if

$$p(A \cap B \mid C) = p(A \mid C)\, p(B \mid C).$$

For the following table of probabilities, verify that the classifications A, $\bar{A}$ and B, $\bar{B}$ are conditionally independent given C, and are also conditionally independent given $\bar{C}$, but are not unconditionally independent.

$$p(A \cap B \cap C) = 0{\cdot}10 \quad p(A \cap B \cap \bar{C}) = 0{\cdot}21$$

$$p(\bar{A} \cap B \cap C) = 0{\cdot}15 \quad p(\bar{A} \cap B \cap \bar{C}) = 0{\cdot}14$$

$$p(A \cap \bar{B} \cap C) = 0{\cdot}06 \quad p(A \cap \bar{B} \cap \bar{C}) = 0{\cdot}15$$

$$p(\bar{A} \cap \bar{B} \cap C) = 0{\cdot}09 \quad p(\bar{A} \cap \bar{B} \cap \bar{C}) = 0{\cdot}10$$

8. Construct a table of probabilities such that the classifications A, $\bar{A}$ and B, $\bar{B}$ are not conditionally independent given C, and are not conditionally independent given $\bar{C}$, but are unconditionally independent.

CHAPTER 2
DISCRETE PROBABILITY DISTRIBUTIONS

§6. Binomial distribution

In order to distinguish between a single experiment, and an experiment which consists of several repetitions, we describe the single experiment as a **trial**. Consider a trial with just two possible outcomes and call them **success** and **failure**. Let θ be the probability of success, whence $1-\theta$ is the probability of failure. Suppose that an experiment consists of n mutually independent trials, so that 2^n outcomes are possible. The probability for each outcome where j successes and $(n-j)$ failures are observed in some order is $\theta^j(1-\theta)^{n-j}$. Now the number of outcomes in the category described is the number of arrangements of j successes and $(n-j)$ failures in a sequence of n trials, namely $n!/j!(n-j)!$. This quantity is known as a binomial coefficient and usually denoted by $\binom{n}{j}$, although older notations such as ${}^nC_j$ are still current. We conclude that the probability for j successes in n trials, say a_j, is given by

$$a_j = \binom{n}{j}\theta^j(1-\theta)^{n-j}.$$

The events which correspond to the values $0, 1, \ldots, n$ of j are mutually exclusive and exhaustive, so that

$$\sum_{j=0}^{n}\binom{n}{j}\theta^j(1-\theta)^{n-j} = 1.$$

This result follows alternatively from the binomial theorem since the left-hand side is the expansion of $\{\theta+(1-\theta)\}^n$.

A sequence of probabilities $\{a_j\}$ exists for every pair (n, θ), where n is a positive integer and θ is a real number such that $0 < \theta < 1$. We say that $\{a_j\}$ is a **binomial distribution** with **index** n and **parameter** θ. The probabilities $a_0, a_1, \ldots, a_n$ steadily increase to a maximum value and then steadily decrease. This is shown by considering the ratios

$$a_j/a_{j-1} = (n-j+1)\theta/j(1-\theta) \quad (j = 1, 2, \ldots, n).$$

The inequality $a_j > a_{j-1}$ is equivalent to $(n-j+1)\theta > j(1-\theta)$, which simplifies to $j < (n+1)\theta$. Similarly $a_{j+1} < a_j$ is equivalent to $(j+1) > (n+1)\theta$. Hence a_j is the maximum probability if $j < (n+1)\theta < j+1$, in which case j is the integral part of $(n+1)\theta$. If $(n+1)\theta$ is an integer, say k, then $a_k = a_{k-1}$. Fig. 3 displays the binomial distribution for $n = 5$ and $\theta = 0{\cdot}3$.

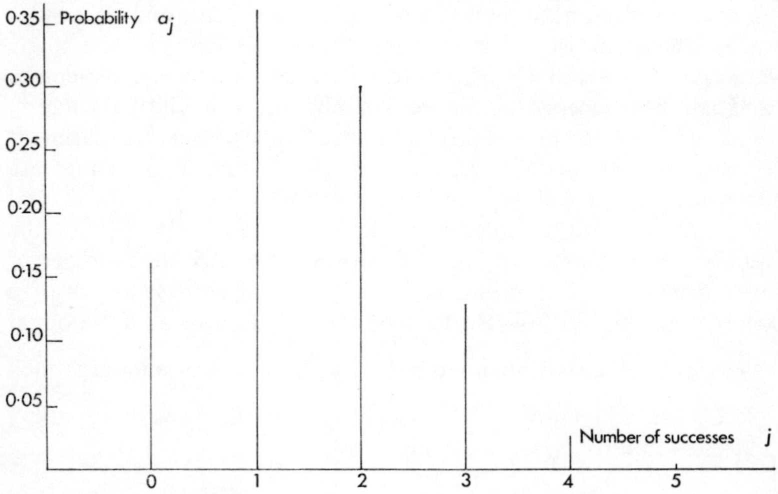

Fig. 3. Binomial distribution with $n = 5$ and $\theta = 0{\cdot}3$.

The binomial distribution is tabled by Pearson and Hartley (1966), and other tables are listed by Greenwood and Hartley (1962). In the absence of a convenient table, the probabilities can be calculated by using the expression for a_j/a_{j-1} as a recurrence formula. We begin with $a_0 = (1-\theta)^n$ and derive $a_1, a_2, \ldots, a_n$ from successive multiplications by the ratios for $j = 1, 2, \ldots, n$ respectively. The condition $\Sigma a_j = 1$ provides a check on the results.

A practical example of the binomial distribution arises in connection with families containing a specified number of children. If the probability of a boy is θ for all births, and the results of successive births are independent events, then the probability for j boys in a family with n children is a_j. Usually θ is unknown, but we suppose that an estimate $\hat{\theta}$ is available calculated from data giving information about r families of size n. Let r_j be the observed frequency of families with j boys. The sequence $\{r_j\}$ is described as a frequency distribution. We can examine whether the theory gives a satisfactory explanation for the data by comparing the estimated probability,

$$\hat{a}_j = \binom{n}{j} \hat{\theta}^j (1-\hat{\theta})^{n-j},$$

with the frequency-ratio r_j/r for $j = 0, 1, \ldots, n$. Alternatively, we can compare the **expected frequency** s_j, defined by

$$s_j = r\hat{a}_j,$$

with the observed frequency r_j for each j. Expected frequencies are seldom integers but they provide a direct comparison with the data as usually presented, and the second method is the one generally followed.

TABLE 1

Frequencies of boys among the first 5 children in 2571 families of size 5 or more

Number of boys	*Observed frequency*	*Expected frequency*
0	63	69·18
1	369	366·91
2	786	778·39
3	821	825·69
4	447	437·93
5	85	92·91
Total	2571	2571·01

Table 1 is based on data compiled by Edwards and Fraccaro (1960) which refer to the offspring of Swedish ministers of religion between 1585 and 1920, concerning whom detailed and accurate information has been recorded. The second column gives the observed frequencies of 0, 1, . . . , 5 boys among the first 5 children in 2571 families of size 5 or more. In the third column are the corresponding expected frequencies obtained by taking $r = 2571$, $n = 5$, and $\hat{\theta} = 0{\cdot}51474$. Here θ is estimated by the observed proportion of boys

$$\{369(1)+786(2)+821(3)+447(4)+85(5)\}/2571(5).$$

We derive the method of estimation from general principles in chapter 11, and explain then how the two sets of frequencies are compared.

§7. Poisson distribution

Consider a binomial distribution with index n and parameter θ. We shall find the limit of each probability a_j when $n \to \infty$ and $\theta \to 0$ in such a way that $n\theta$ is fixed. Let the fixed value be λ, where λ is necessarily positive. Then

$$\binom{n}{j}\theta^j(1-\theta)^{n-j} = \left\{\frac{n(n-1)\ldots(n-j+1)}{(n-\lambda)^j}\right\}\frac{\lambda^j}{j!}\left(1-\frac{\lambda}{n}\right)^n.$$

As $n \to \infty$, the first term on the right tends to 1 and the last to $e^{-\lambda}$. Hence

$$\lim_{n\to\infty}\binom{n}{j}\theta^j(1-\theta)^{n-j} = \frac{\lambda^j}{j!}e^{-\lambda}.$$

Denote $\lambda^j e^{-\lambda}/j!$ by b_j $(j = 0, 1, 2, \ldots)$. For each n and θ, the probabilities $a_0, a_1, \ldots, a_n$ satisfy $\Sigma a_j = 1$, which suggests that $\Sigma b_j = 1$ also. The result is checked by evaluating

$$\sum_{j=0}^{\infty}\frac{\lambda^j}{j!}e^{-\lambda} = e^{-\lambda}e^{\lambda} = 1.$$

A sequence $\{b_j\}$ exists for every positive real number λ. We say that $\{b_j\}$ is a Poisson distribution with parameter λ. Tables of the distribution are given by Pearson and Hartley (1966). The probabilities can be calculated recursively by means of the ratios

$$b_j/b_{j-1} = \lambda/j \quad (j = 1, 2, 3, \ldots).$$

Starting with $b_0 = e^{-\lambda}$, we derive $b_1, b_2, \ldots$ from successive multiplications by $\lambda, \lambda/2, \lambda/3, \ldots$ respectively. This procedure also shows that for $\lambda < 1$ the probabilities steadily decrease, while for $\lambda > 1$ they steadily increase to a maximum and then steadily decrease. Fig. 4 displays the Poisson distribution for $\lambda = 1{\cdot}5$.

The conditions under which the distribution has been derived are likely to be realized, for example, during the inspection of a batch of

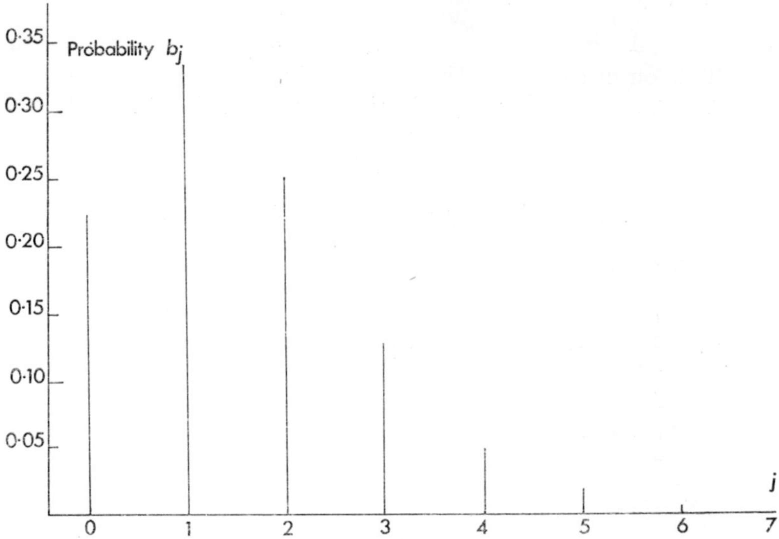

Fig. 4. Poisson distribution with $\lambda = 1·5$.

mass-produced components. When the number in the batch is large and there is a small probability that an individual component is defective, the probability that the batch contains j defective components is approximately b_j for some λ. Table 2 compares the binomial probabilities $\{a_j\}$ for $n = 60$ and $\theta = 0·04$ with the Poisson probabilities $\{b_j\}$ for $\lambda = 2·4$.

TABLE 2

Binomial probabilities for $n = 60$, $\theta = 0·04$
and Poisson probabilities for $\lambda = 2·4$

	0	1	2	3	4	5	6 *or more*
Binomial	0·086	0·216	0·265	0·214	0·127	0·059	0·033
Poisson	0·091	0·218	0·261	0·209	0·125	0·060	0·036

For a practical example, consider an experiment performed by 'Student' (1907). A liquid containing yeast cells was mixed well and spread out in a layer 0·01 mm. thick over 400 units of area, the unit area being 1/400 sq. mm. The total number of cells in such an experiment is large, about 500, and the probability that a particular unit area is occupied by a particular cell is small, about 1/400. Consequently the probability is approximately b_j that a particular unit area contains j cells, and the frequencies for 400 unit areas should conform to a Poisson distribution. The second column of Table 3 gives for $j = 0, 1, 2, \ldots$ the frequency of unit areas containing j cells. In the

TABLE 3

Frequencies of yeast cells in 400 unit areas

Number of cells	*Observed frequency*	*Expected frequency*
0	103	106·59
1	143	140·96
2	98	93·21
3	42	41·09
4	8	13·59
5	4	3·59
6 or more	2	0·97
Total	400	400·00

third column are the corresponding expected frequencies, which are calculated from

$$400\hat{\lambda}^{j} e^{-\hat{\lambda}}/j! \quad (j = 0, 1, 2, \ldots),$$

where $\hat{\lambda} = 1{\cdot}3225$. This value is the average number of cells per unit area, namely

$$\{143(1)+98(2)+42(3)+8(4)+4(5)+2(6)\}/400.$$

As before, we defer comment on the method of estimating λ and the agreement between the two sets of frequencies.

§8. Hypergeometric distribution

Consider an electoral constituency which contains t voters, of whom r are Conservative and $(t-r)$ are Labour. A group of n voters is selected. Enquiry shows that j are Conservative and $(n-j)$ are Labour. We suppose that the method by which the group is chosen is such that all possible groups of size n have the same probability of selection. There are $\binom{t}{n}$ different groups, and the probability in question is therefore $1/\binom{t}{n}$. However, j Conservative voters can be selected from a total of r in $\binom{r}{j}$ ways, and similarly $(n-j)$ Labour voters can be selected in $\binom{t-r}{n-j}$ ways. Let c_j be the probability that the group selected contains j Conservative voters. We conclude that

$$c_j = \binom{r}{j}\binom{t-r}{n-j}\bigg/\binom{t}{n}.$$

The value of j must satisfy

$$0 \leqslant j \leqslant r \quad \text{and} \quad 0 \leqslant n-j \leqslant t-r.$$

Thus the probabilities $\{c_j\}$ are non-zero when

$$\max(0, r+n-t) \leqslant j \leqslant \min(r, n).$$

The non-zero probabilities correspond to mutually exclusive and exhaustive events, so that

$$\Sigma \binom{r}{j}\binom{t-r}{n-j}\bigg/\binom{t}{n} = 1,$$

where the summation is taken over the range of j indicated. We say that $\{c_j\}$ is a **hypergeometric** distribution. The name arises from the appearance of such quantities as coefficients in one of the series solutions of the hypergeometric equation.

In the example chosen to introduce the distribution, the size of the group selected is usually small compared with the total number of voters. On the other hand, the numbers of Conservative and Labour voters in the constituency have the same order of magnitude. Consider the limit of c_j when n is fixed, $t \to \infty$, and $r/t \to \theta$. The version

$$c_j = \binom{n}{j}\binom{t-n}{r-j}\bigg/\binom{t}{r}$$

is obtained by expansion of the binomial coefficients and rearrangement. Here $\binom{n}{j}$ is fixed and

$$\binom{t-n}{r-j}\bigg/\binom{t}{r}$$

$$= \frac{[r(r-1)\ldots(r-j+1)]\,[(t-r)(t-r-1)\ldots\{(t-r)-(n-j)+1\}]}{t(t-1)\ldots(t-n+1)}.$$

The ratio is asymptotically equal to $r^j(t-r)^{n-j}/t^n$ and its limiting value is $\theta^j(1-\theta)^{n-j}$. Under the conditions specified, we therefore find that

$$c_j \to \binom{n}{j} \theta^j(1-\theta)^{n-j}.$$

Thus the limiting distribution is binomial with index n and parameter θ.

§9. Discrete variates

A **variate** or **random variable** X is a function on the outcome space. We define X by assigning one member of a sequence of distinct real numbers $x_0, x_1, x_2, \ldots$ to each outcome.

Denote the outcomes by $\omega_0, \omega_1, \omega_2, \ldots$ and the corresponding probabilities by $p_0, p_1, p_2, \ldots$ respectively. The axioms imply that

$$p_i \geqslant 0 \text{ and } \Sigma p_i = 1.$$

Any sequence of probabilities $\{p_i\}$ which satisfies these two conditions is known as a **discrete** probability distribution. Let $\omega^x_{j0}, \omega^x_{j1}, \omega^x_{j2}, \ldots$ be the outcomes which are assigned the value x_j, and write f_j for the sum of the corresponding probabilities. Thus

$$p(X = x_j) = f_j \quad \text{where} \quad f_j = p^x_{j0}+p^x_{j1}+p^x_{j2}+\ldots.$$

From the definition of f_j, and the fact that each p_i appears in just one f_j, we obtain

$$f_j \geqslant 0 \quad \text{and} \quad \Sigma f_j = 1.$$

Consequently $\{f_j\}$ is a discrete probability distribution. We say that X is a **discrete variate** with the distribution $\{f_j\}$. In most practical applications, x_j is a multiple of j and the sequence $\{f_j\}$ is defined by a single mathematical function which provides a convenient summary of the distribution.

Example 1. An electoral constituency of size t contains r Conservative voters and $(t-r)$ Labour voters. Let X be the number of Conservative voters in a group of size n, so that $X = j$ for $\binom{r}{j}\binom{t-j}{n-r}$ different outcomes. When the $\binom{t}{n}$ groups of size n have the same probability of selection,

$$p(X = j) = \binom{r}{j}\binom{t-r}{n-j}\Big/\binom{t}{n}.$$

These ideas can be extended as follows. Suppose that $y_0, y_1, y_2, \ldots$ is another sequence of distinct real numbers, and assign one member of the sequence to each outcome. Let $\omega^y_{k0}, \omega^y_{k1}, \omega^y_{k2}, \ldots$ be the outcomes which are assigned the value y_k, and define the variate Y as the function which takes the value y_k on each of these outcomes. Write

$$g_k = p^y_{k0}+p^y_{k1}+p^y_{k2}+\ldots.$$

The probability distribution of Y is then given by

$$p(Y = y_k) = g_k \quad (k = 0, 1, 2, \ldots).$$

We can also enumerate the outcomes $\omega_{jk0}, \omega_{jk1}, \omega_{jk2}, \ldots$ which have been assigned the pair of values (x_j, y_k). Let h_{jk} be the sum of the corresponding probabilities. By the same argument as for $\{f_j\}$,

$$h_{jk} \geqslant 0 \quad \text{and} \quad \Sigma\Sigma h_{jk} = 1.$$

The **joint** or **bivariate** probability distribution of X and Y is defined by

$$p(X = x_j, Y = y_k) = h_{jk} \quad (j, k = 0, 1, 2, \ldots).$$

Example 2. An electoral constituency of size t contains r Conservative, s Labour, and $t-r-s$ Liberal voters. Let X be the number of Conservative voters and Y the number of Labour voters in a group of size n. Thus $X = j$ and $Y = k$ for all groups which contain j Conservative, k Labour, and $n-j-k$ Liberal voters. There are

$$\binom{r}{j}\binom{s}{k}\binom{t-r-s}{n-j-k}$$

such groups. If each group of size n has the same probability of selection, then

$$p(X = j, Y = k) = \binom{r}{j}\binom{s}{k}\binom{t-r-s}{n-j-k}\bigg/\binom{t}{n}.$$

Consider the set of outcomes assigned the pair of values (x_j, y_k). The aggregate of such sets for $k = 0, 1, 2, \ldots$ is the set of outcomes assigned the value x_j. Consequently

$$\Sigma_k\, h_{jk} = f_j,$$

and similarly

$$\Sigma_j\, h_{jk} = g_k.$$

Thus the distributions of X and Y individually can be derived from their joint distribution. We can display the probabilities $\{h_{jk}\}$ in a two-dimensional array with h_{jk} in the cell formed by the jth row and kth column. The distributions $\{f_j\}$ and $\{g_k\}$ are then the row and column totals of the array, and for this reason they are known as **marginal** distributions.

For any y_k such that $p(Y = y_k) > 0$, we have

$$p(X = x_j \mid Y = y_k) = h_{jk}/g_k.$$

When k is fixed and $j = 0, 1, 2, \ldots$, a sequence of conditional probabilities is obtained, such that

$$h_{jk}/g_k \geqslant 0 \quad \text{and} \quad \Sigma_j \, h_{jk}/g_k = 1.$$

Thus $\{h_{jk}/g_k\}$ is a discrete probability distribution. We say that $p(X = x_j \mid Y = y_k)$ defines the probability distribution of X **conditional** on $Y = y_k$. Similarly, $p(Y = y_k \mid X = x_j)$ defines the probability distribution of Y conditional on $X = x_j$ whenever $f_j > 0$.

The variates X and Y are **independent** if

$$h_{jk} = f_j \, g_k \text{ for all } j \text{ and } k.$$

Independence implies that

$$p(X = x_j \mid Y = y_k) = p(X = x_j)$$

for all x_j and y_k such that $p(Y = y_k) > 0$. A similar result holds with the variates interchanged.

Example 3. Consider the situation described in the previous example. We have found the marginal distribution of X in example 1, because the j Conservative and $(n-j)$ non-Conservative voters in a group of size n are selected from r Conservative and $(t-r)$ non-Conservative voters in the constituency. Similarly

$$p(Y = k) = \binom{s}{k}\binom{t-s}{n-k}\bigg/\binom{t}{n}.$$

The distribution of Y conditional on $X = j$ is

$$p(Y = k \mid X = j) = \binom{s}{k}\binom{t-r-s}{n-j-k}\bigg/\binom{t-r}{n-j}.$$

Both the marginal and conditional distributions of Y are hypergeometric, but they are distinct, so that X and Y are not independent.

Example 4. The random decimal digit generator ERNIE is used to make monthly prize draws for Premium Savings Bonds. It is constructed so as to assign a probability of 1/10 to each of the digits 0, 1, . . . , 9. If two such generators are operated simultaneously, then the probability that a particular pair of digits is observed during the same time-interval is 1/100, on the assumption that the individual digits X and Y are independent variates. Thus, in two parallel sequences of 10,000 digits each, the expected frequency for each pair is 100. In Table 4, which is taken from Thomson (1959), rows are

TABLE 4

Frequencies of digital pairs from two generators of random numbers

j \ k	9	8	7	6	5	4	3	2	1	0
9	97	94	93	107	90	95	100	102	92	100
8	87	100	100	105	100	89	98	101	112	104
7	90	81	99	107	94	113	93	95	100	100
6	99	96	102	105	111	121	95	103	107	97
5	98	103	95	97	91	96	101	97	97	96
4	95	97	110	82	98	110	106	112	100	116
3	103	97	100	104	79	89	100	106	100	106
2	94	115	91	109	101	107	87	107	105	104
1	110	96	83	120	111	90	111	98	96	97
0	97	117	99	100	96	116	93	99	103	93

labelled by the values of X ($j = 0, 1, 2, \ldots, 9$) and columns by the values of Y ($k = 0, 1, \ldots, 9$). An inspection of the frequencies appears to confirm that X and Y are independent variates.

§10. Discrete multivariate distributions

In general, any finite set of variates $X, Y, \ldots, Z$ can be defined on the same outcome space. A straightforward extension of the method for deriving the bivariate distribution of X and Y gives the total probability $h_{jk\ldots m}$ for all outcomes which have been assigned the set of values $(x_j, y_k, \ldots, z_m)$. The **joint** or **multivariate** distribution of $X, Y, \ldots, Z$ is defined by

$$p(X = x_j, Y = y_k, \ldots, Z = z_m) = h_{jk\ldots m},$$

where $h_{jk\ldots m} \geqslant 0$ and $\Sigma_{jk\ldots m}\, h_{jk\ldots m} = 1$. The marginal distribution of any set of variates is obtained by summation over the values of the remainder. In particular,

$$p(X = x_j) = \Sigma_{k\ldots m}\, h_{jk\ldots m}.$$

We say that $X, Y, \ldots, Z$ are **mutually independent** if

$$p(X=x_j, Y=y_k, \ldots, Z=z_m) = p(X=x_j)p(Y=y_k) \ldots p(Z=z_m) \quad \text{for all } x_j, y_k, \ldots, z_m.$$

Example 1. An experiment consists of n independent trials, which are divided into successive groups with sizes $n_1, n_2, \ldots, n_t$ where $\Sigma n_i = n$. The outcome of each trial is either success or failure, which occur with probabilities θ and $1-\theta$ respectively. Let X_i be the frequency of success in the n_i trials of the ith group ($i = 1, 2, \ldots, t$). The probability for any sequence of trials such that $X_1 = x_1$, $X_2 = x_2, \ldots, X_t = x_t$ is $\theta^x(1-\theta)^{n-x}$, where $x = \Sigma x_i$. Since the x_i successes in the ith group can be assigned to the n_i trials in $\binom{n_i}{x_i}$ ways, the total number of experimental outcomes with $X_i = x_i$ for all i is $\prod\binom{n_i}{x_i}$. Thus

$$p(X_1 = x_1, X_2 = x_2, \ldots, X_t = x_t) = \prod_{i=1}^{t} \binom{n_i}{x_i} \theta^{x_i} (1-\theta)^{n_i - x_i}.$$

The marginal distribution of X_i is binomial with index n_i and parameter θ. We therefore conclude that the variates $X_1, X_2, \ldots, X_t$ are mutually independent.

Example 2. An experiment consists of n independent trials, at each of which the possible outcomes are $\omega_1, \omega_2, \ldots, \omega_t$. Denote the corresponding probabilities by $\theta_1, \theta_2, \ldots, \theta_t$ and the frequencies with which the outcomes occur by $X_1, X_2, \ldots, X_t$ respectively. Here

$$\Sigma\theta_i = 1 \quad \text{and} \quad \Sigma X_i = n.$$

Any sequence of trials for which $X_1 = x_1, X_2 = x_2, \ldots, X_t = x_t$ has probability $\theta_1^{x_1} \theta_2^{x_2} \ldots \theta_t^{x_t}$. The number of such sequences is

$$\binom{n}{x_1}\binom{n-x_1}{x_2}\binom{n-x_1-x_2}{x_3}\cdots\binom{n-x_1-x_2-\ldots-x_{t-1}}{x_t},$$

because the first term is the number of ways of selecting x_1 trials for ω_1 out of n trials, the second is the number of ways of selecting x_2 trials for ω_2 out of the remaining $(n-x_1)$ trials, $\ldots$, and the last gives the single way of selecting the trials for ω_t when those for $\omega_1, \omega_2, \ldots, \omega_{t-1}$ are determined. The product of binomial coefficients simplifies to $n!/x_1!x_2! \ldots x_t!$, known as a multinomial coefficient. Hence the joint probability distribution of $X_1, X_2, \ldots, X_t$ is given by

$$p(X_1=x_1, X_2=x_2, \ldots, X_t=x_t) = n!\theta_1^{x_1}\,\theta_2^{x_2}\ldots\theta_t^{x_t}/x_1!x_2!\ldots x_t!.$$

We say that $X_1, X_2, \ldots, X_t$ have a **multinomial** distribution with index n and parameters $\theta_1, \theta_2, \ldots, \theta_t$. The marginal distribution of X_i is obtained by classifying the outcome of each trial as success when ω_i occurs and failure otherwise. This shows that X_i has a binomial distribution with index n and parameter θ_i $(i = 1, 2, \ldots, t)$. The variates $X_1, X_2, \ldots, X_t$ are not independent, because their joint distribution is different from the product of the marginal distributions.

In practice, the outcomes $\omega_1, \omega_2, \ldots, \omega_t$ are often associated with the values taken by a discrete variate. Suppose, for instance, that n families of size 5 are classified into X_1 with 0 boys, X_2 with 1 boy, ..., and X_6 with 5 boys. The probability that a family of size 5 contains j boys is

$$\theta_{j+1} = \binom{5}{j}\theta^j(1-\theta)^{5-j} \quad (j = 0, 1, \ldots, 5),$$

where θ is the probability for a boy. Therefore $X_1, X_2, \ldots, X_6$ have a multinomial distribution with index n and parameters $\theta_1, \theta_2, \ldots, \theta_6$. The frequencies observed in the $n = 2571$ Swedish families considered in §**6** were

$$x_1 = 63,\ x_2 = 369,\ x_3 = 786,\ x_4 = 821,\ x_5 = 447,\ x_6 = 85.$$

Exercises 2

1. The frequency distribution of the number of boys in the first 4 children of 3343 Swedish families with 4 or more children is as follows (Edwards and Fraccaro, 1960). Calculate the expected frequencies on the assumption that the data follow a binomial distribution.

Number of boys	0	1	2	3	4
Frequency	183	789	1250	875	246

2. Let X_n be the number of successes in n independent trials with probability θ for success. Use integration by parts to prove that

$$p(X_n \leqslant j) = \frac{n!}{j!(n-j-1)!}\int_\theta^1 x^j(1-x)^{n-j-1}\,dx.$$

3. A sequence of independent trials with probability θ for success is continued until the mth success occurs. Let Y_m be the number of trials required. Prove that

$$p(Y_m = m+k) = \binom{m+k-1}{m-1} \theta^m (1-\theta)^k \ (k = 0, 1, 2, \ldots).$$

Write d_k for the probability on the right. Show that the probabilities $\{d_k\}$ are obtained as terms in the expansion of $\{1/\theta-(1-\theta)/\theta\}^{-m}$. We say that $\{d_k\}$ is a **negative binomial** distribution.

4. Let X_n and Y_m be the variates defined in the two previous exercises. Show that the event $X_{m+t-1} \leqslant m-1$ contains the same outcomes as the event $Y_m \geqslant m+t$. Deduce the identity

$$\sum_{j=0}^{m-1} \binom{m+t-1}{j} \theta^j (1-\theta)^{m+t-j-1} \equiv \sum_{k=t}^{\infty} \binom{m+k-1}{k} \theta^m (1-\theta)^k.$$

5. A record was made of the number of times that each member of a fleet of army vehicles entered the workshop for repair in a year. The frequency distribution of repair visits is taken from Drinkwater and Hastings (1967) and given below. Calculate the expected frequencies on the assumption that the data follow a Poisson distribution.

Number of visits	0	1	2	3	4	5
Frequency	295	190	53	5	5	2

6. X has a Poisson distribution with parameter λ. Prove that

$$p(X \leqslant j) = \frac{1}{j!}\int_{\lambda}^{\infty} x^j e^{-x}\, dx.$$

7. An experiment consists of n independent trials with probability θ for success. Let X be the total number of successes and Y the number of successes in the first m trials ($m < n$). Show that

$$p(Y = k \mid X = j) = \binom{m}{k}\binom{n-m}{j-k}\Big/\binom{n}{j}.$$

8. A room contains r particles of dust. Consider a unit of volume so small that at most one particle may be contained in it. Expressed in terms of this unit, the volume of the room is t, and the volume of a particular region in the room is n. Suppose that the particles can be distinguished from each other, and that all spatial configurations of

them are equally probable. Let c_j be the probability that the region contains j particles. Show that

$$c_j = \binom{r}{j}\binom{t-r}{n-j}\bigg/\binom{t}{n}.$$

9. The magnitude of the quantities in the previous exercise is such that r, n and t are all large but $r \ll t$ and $n \ll t$. Denote rn/t by λ. When λ is fixed and r, n, t all tend to infinity such that $r/t \to 0$ and $n/t \to 0$, prove that $c_j \to \lambda^j e^{-\lambda}/j!$.

10. Suppose that $X_1, X_2, \ldots, X_t$ have a multinomial distribution with index n and parameters $\theta_1, \theta_2, \ldots, \theta_t$. Write

$$n\theta_1 = \lambda_1, n\theta_2 = \lambda_2, \ldots, n\theta_{t-1} = \lambda_{t-1}.$$

For fixed values of $\lambda_1, \lambda_2, \ldots, \lambda_{t-1}$, prove that

$$\lim_{n \to \infty} p(X_1 = x_1, X_2 = x_2, \ldots, X_t = x_t) = \prod_{i=1}^{t-1} \lambda_i^{x_i} e^{-\lambda_i}/x_i!.$$

CHAPTER 3

CONTINUOUS PROBABILITY DISTRIBUTIONS

§11. Continuous variates

Many experiments are concerned with measurements of physical quantities such as length, time, current or temperature. The system of outcomes then usually consists of equal intervals on a continuous scale of measurement. For example, if the heights of male students on entering a University were recorded to the nearest cm., each interval would take the form $(h \pm \frac{1}{2})$ cm., where h is an integer, and similarly for any other unit of length. In general, by the substitution of a more precise measuring instrument, each outcome can be subdivided, thus forming a new system of outcomes. There will be a limit to how far the subdivision can be continued in practice, but if it were continued until measurements were made with infinite precision, then the corresponding space would consist of every point in a particular range of the measurement scale. We suppose in what follows that the outcome space for individual measurements is a straight line, and assign zero probability to impossible events such as an interval of negative heights.

A variate X is again defined as a function on the outcome space. The probability of the event $X \leqslant x$ is denoted by $F(x)$ and known as the **distribution function** of X. Thus

$$p(X \leqslant x) = F(x).$$

When $x \to \infty$, all outcomes are successively contained in $X \leqslant x$, and $p(X \leqslant x) \to 1$. When $x \to -\infty$, all outcomes are successively excluded from $X \leqslant x$, and $p(X \leqslant x) \to 0$. Consequently

$$\lim_{x \to \infty} F(x) = 1 \quad \text{and} \quad \lim_{x \to -\infty} F(x) = 0.$$

For $a < b$, the events $X \leqslant a$ and $a < X \leqslant b$ are mutually exclusive, and their union is $X \leqslant b$. Hence

$$p(X \leqslant b) = p(X \leqslant a) + p(a < X \leqslant b).$$

Since $p(a < X \leqslant b) \geqslant 0$, we obtain $F(b) - F(a) \geqslant 0$, so that $F(x)$ is a non-decreasing function.

If the possible values of X are $x_0, x_1, x_2, \ldots$ and $p(X = x_j) = f_j$, then X is a discrete variate with the distribution $\{f_j\}$. In this case

$$F(x) = \Sigma f_j,$$

where the summation is taken over all values of j such that $x_j \leqslant x$.

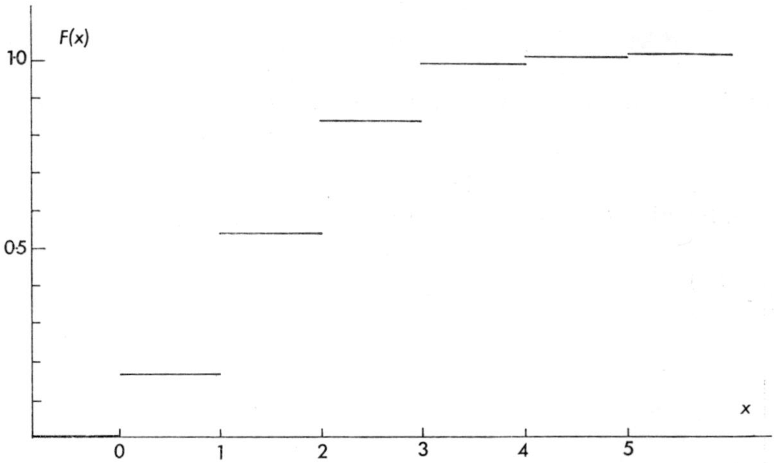

Fig. 5. Distribution function for binomial distribution with $n = 5$ and $\theta = 0{\cdot}3$.

Thus $F(x)$ is constant in each interval of the form $x_j \leqslant x < x_{j+1}$ and increases by f_j at $x = x_j$ for each j. Fig. 5 displays $F(x)$ when

$$p(X = j) = \binom{5}{j} (0{\cdot}3)^j \, (0{\cdot}7)^{5-j} \; (j = 0, 1, \ldots, 5).$$

There is also the possibility that $F(x)$ increases gradually and not by finite steps. We therefore pass from sums to integrals and suppose that a continuous function $f(x)$ exists such that

$$F(x) = \int_{-\infty}^{x} f(t)\,dt.$$

By a theorem of the integral calculus, $f(x)$ is the derivative of $F(x)$, and since $F(x)$ is non-decreasing, $f(x) \geqslant 0$. The condition $F(-\infty)=0$ is satisfied, and $F(\infty) = 1$ implies that

$$\int_{-\infty}^{\infty} f(x)\,dx = 1.$$

We now have

$$p(a < X \leqslant b) = \int_{a}^{b} f(x)\,dx.$$

Fix b and let $a \to b$ from below. The limit of the integral is zero, so that $p(X = x) = 0$ for all x. Thus $p(a < X \leqslant b)$ and $p(a < X < b)$ are identical. When $a = x$ and $b = x+dx$, we obtain

$$p(x < X < x+dx) = f(x)\,dx,$$

which is known as a **probability element**.

A function $f(x)$ with the properties

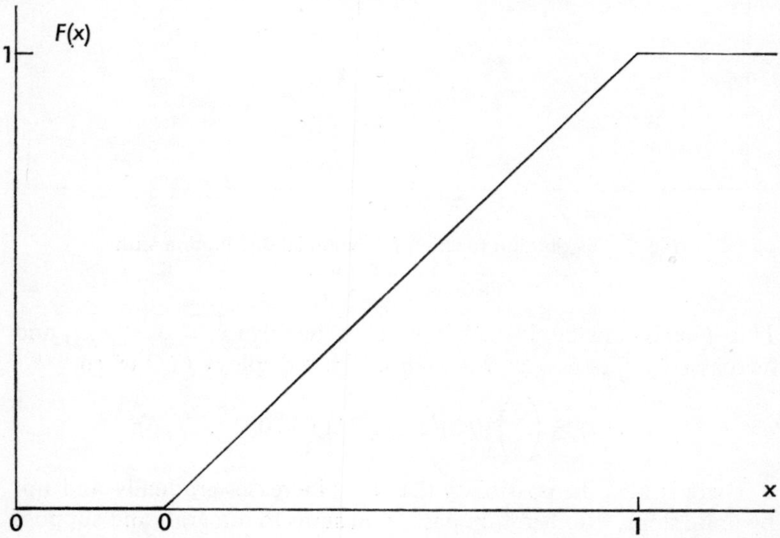

Fig. 6. Distribution function for continuous uniform distribution.

$$f(x) \geqslant 0 \quad \text{and} \quad \int_{-\infty}^{\infty} f(x)\,dx = 1$$

defines a **continuous** probability distribution. We say that X is a **continuous variate** with the **density function** $f(x)$. For example, define

$$f(x) = \begin{cases} 1 & (0 < x < 1), \\ 0 & \text{otherwise.} \end{cases}$$

Both conditions for a density function are satisfied, and X is said to have a **uniform** distribution on the interval (0, 1). Fig. 6 displays the corresponding distribution function, which is given by

$$F(x) = \begin{cases} 0 & (x \leqslant 0) \\ x & (0 < x < 1) \\ 1 & (x \geqslant 1). \end{cases}$$

The choice of the term 'density' arises from the physical analogy of a thin rod with unit total mass where $f(x)$ represents the density of mass at the point x. Many functions which satisfy the conditions on $f(x)$ occur in statistical methodology. Two more examples are given in the following sections, and others will arise naturally as the subject is developed.

§12. Exponential distribution

Suppose that

$$f(x) = \begin{cases} \lambda e^{-\lambda x} & (x > 0; \lambda > 0) \\ 0 & (x \leqslant 0). \end{cases}$$

The conditions for a density function are satisfied, because $f(x) \geqslant 0$ and

$$\int_{0}^{\infty} \lambda e^{-\lambda x}\,dx = [-e^{-\lambda x}]_0^{\infty} = 1.$$

We say that X has an **exponential** distribution with parameter λ.

The distribution function of X is

$$F(x) = \begin{cases} 1 - e^{-\lambda x} & (x > 0) \\ 0 & (x \leqslant 0), \end{cases}$$

and hence when $0 < a < b$

$$p(a < X < b) = e^{-\lambda a} - e^{-\lambda b}.$$

These expressions show that the dimension of λ must be the reciprocal of the dimension of x because probability, and therefore $e^{-\lambda x}$, is a dimensionless quantity.

An argument leading to the exponential distribution is the following. Consider an incident which occurs randomly at instants of time, such as a road accident or the emission of an α-particle by a radioactive source. Let x be the interval between successive occurrences of the incident, and write

$$p(X > x) = \mathscr{F}(x).$$

Then $\mathscr{F}(x)$ is also the probability that no incidents occur in an interval of duration x beginning just after the occurrence of an incident. Subdivide this interval into two successive intervals of lengths r and s, where $x = r+s$. For convenience, refer to each of the three intervals by its length. Define the outcomes of each interval by the number of incidents which are observed. We make the assumption that the experiment which consists of observing the number of incidents in r is independent of the corresponding experiment for s, whatever the values of r and s. Let A denote the outcome that no incidents occur in r and let B denote the corresponding outcome for s. If A is followed by B, then no incidents occur in x. Therefore

$$\mathscr{F}(r+s) = \mathscr{F}(r)\mathscr{F}(s) \quad \text{for all } r \text{ and } s.$$

Differentiate with respect to r, whence

$$\mathscr{F}'(r+s) = \mathscr{F}'(r)\mathscr{F}(s),$$

and similarly

$$\mathscr{F}'(r+s) = \mathscr{F}(r)\mathscr{F}'(s).$$

In this way, we obtain

$$\mathscr{F}'(r)/\mathscr{F}(r) = \mathscr{F}'(s)/\mathscr{F}(s).$$

Thus, for some constant λ,

$$\mathscr{F}'(x)/\mathscr{F}(x) = -\lambda.$$

Integration with respect to x gives

$$\mathscr{F}(x) = e^{-\lambda x - \mu},$$

where μ is another constant. Now $F(x) = 1-\mathscr{F}(x)$. Hence λ is positive because $F(\infty) = 1$ and $\mu = 0$ because $F(0) = 0$. We conclude that the distribution of X is exponential with parameter λ.

Experiments on the emission of α-particles by uranium provide data which appear to conform to an exponential distribution, and which thus lend support to the assumption of independence made above. Rutherford and Geiger recorded the duration of 7563 intervals between emission, and their experimental results are presented in Table 5, which is taken from Marsden and Barratt (1911).

TABLE 5

Intervals between emission of α-particles by uranium

Duration of interval (sec.)	*Observed frequency*	*Expected frequency*
0–	3106	3084·38
1–	1763	1826·49
2–	1115	1081·60
3–	658	640·50
4–	389	379·29
5–	206	224·60
6–	130	133·01
7–	86	78·76
8–	42	46·64
9–∞	68	67·72
Total	7563	7562·99

When measurements are made on a continuous scale, it is necessary to collect them into groups so that observed and expected frequencies can be compared. Here the first group contains all intervals of duration less than 1 sec., the second group all intervals of duration greater than 1 sec. but less than 2 sec., and groups which span 1 sec. are used until the last group is reached, containing all intervals of duration greater than 9 sec. The observed frequency distribution corresponding to this grouping is given in the second column of Table 5. An estimate of λ is required in order to calculate the expected frequencies. The dimension of λ is the reciprocal of time, and we explain in chapter 11 why λ is estimated by 0·52395 sec.$^{-1}$. For each relevant pair of values of a and b, the third column of Table 5 contains the expected frequency of intervals with duration greater than a but less than b, namely

$$7563\,\{\exp(-0{\cdot}52395a)-\exp(-0{\cdot}52395b)\}.$$

Agreement between the second and third columns is satisfactory, but more precise comment is deferred until chapter 11, when we examine what differences can reasonably be allowed.

§13. Normal distribution

Consider the function ϕ which is defined by

$$\phi(t) = (2\pi)^{-\frac{1}{2}} \exp(-\tfrac{1}{2}t^2) \quad (-\infty < t < \infty).$$

Then $\phi(t)$ is a positive function which attains its maximum at $t = 0$, and $\phi(t) = \phi(-t)$. Fig. 7 shows that the graph of $\phi(t)$ is a bell-

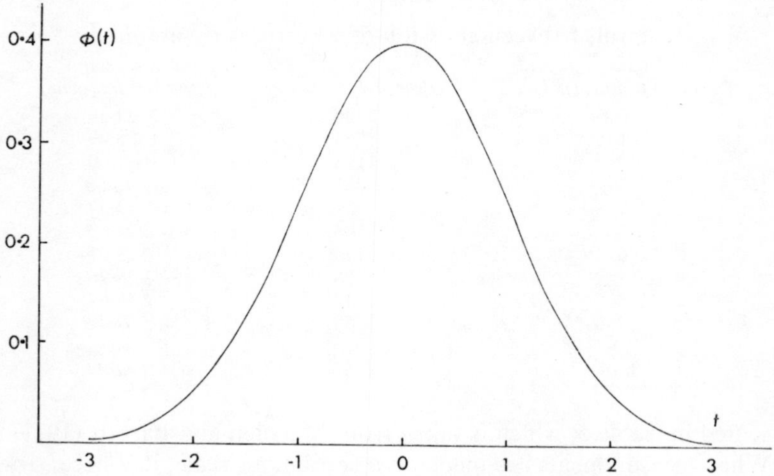

Fig. 7. Standard normal density function.

shaped curve which extends to $t = \pm 3$ approximately before becoming indistinguishable from the t-axis.

Denote the total integral of $\phi(t)$ by I. Then

$$I^2 = \{\int_{-\infty}^{\infty} (2\pi)^{-\frac{1}{2}} \exp(-\tfrac{1}{2}t^2)\, dt\}^2$$

$$= \int_{-\infty}^{\infty} \int_{-\infty}^{\infty} (2\pi)^{-1} \exp\{-\tfrac{1}{2}(t^2+u^2)\}\, dt\, du.$$

The double integral is evaluated by changing from cartesian to polar coordinates. Write

$$t = r \sin\theta \quad \text{and} \quad u = r\cos\theta.$$

An element of area in the new coordinate system is $dr\,(rd\theta)$. Hence

$$I^2 = \int_0^{2\pi} d\theta \int_0^{\infty} (2\pi)^{-1} \exp(-\tfrac{1}{2}r^2)\, r\, dr = 1.$$

Now $I > 0$ and therefore $I = 1$. Since $\phi > 0$ and $I = 1$, we conclude that ϕ is a density function.

The variate T with density function ϕ is said to have a **standard normal** distribution. Denote the distribution function of T by $\Phi(t)$. Thus

$$\Phi(t) = \int_{-\infty}^{t} \phi(u)\, du.$$

We deduce from the property $\phi(u) = \phi(-u)$ that

$$\int_{-\infty}^{-t} \phi(u)\, du = \int_{t}^{\infty} \phi(u)\, du.$$

The right-hand side is $1-\Phi(t)$, and therefore

$$\Phi(-t) = 1-\Phi(t).$$

A graph of $\Phi(t)$ is drawn in Fig. 8. The results concerning ϕ and Φ

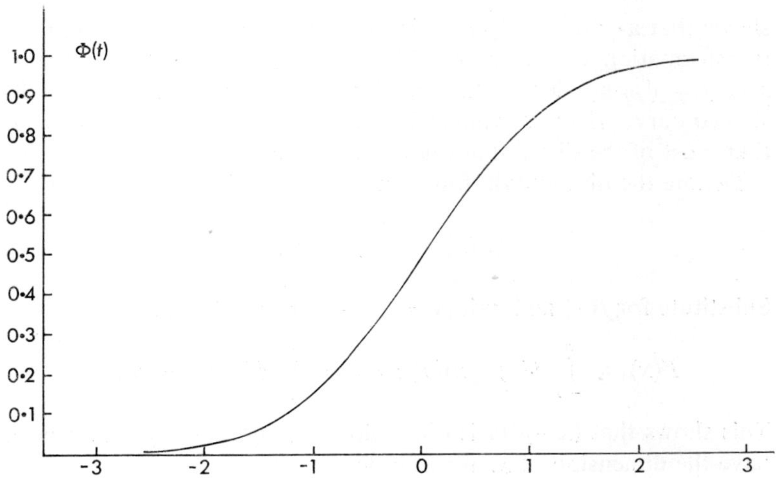

Fig. 8. Standard normal distribution function.

show that it is sufficient to tabulate these functions for $t \geqslant 0$. Tables are given by Pearson and Hartley (1966).

We have proved that

$$\int_{-\infty}^{\infty} (2\pi)^{-\frac{1}{2}} \exp\left(-\tfrac{1}{2}t^2\right) dt = 1.$$

Make the substitution

$$t = (x-\mu)/\sigma,$$

where $-\infty < \mu < \infty$ and $\sigma > 0$. We obtain

$$\int_{-\infty}^{\infty} (2\pi\sigma^2)^{-\frac{1}{2}} \exp\{-(x-\mu)^2/2\sigma^2\}\, dx = 1.$$

Denote the integrand by $f(x)$, so that

$$f(x) = (2\pi\sigma^2)^{-\frac{1}{2}} \exp\{-(x-\mu)^2/2\sigma^2\}.$$

Then $f(x)$ is a density function because it is positive for all x and its total integral is unity.

The variate X with density function $f(x)$ is said to have a **normal** distribution with parameters μ and σ^2. When $\mu = 0$ and $\sigma = 1$, X has a standard normal distribution. The relationship

$$f(x) = \phi\{(x-\mu)/\sigma\}/\sigma$$

shows that a graph of $f(x)$ is obtained from one of $\phi(t)$ by a linear transformation $x = \mu + t\sigma$ applied to the axis of t, and a change of scale $f = \phi/\sigma$ applied to the axis of ϕ. Thus a graph of $f(x)$ is a bell-shaped curve which is symmetrical about the value $x = \mu$ and such that most of the distribution lies in the range $\mu \pm 3\sigma$.

Denote the distribution function of X by $F(x)$, so that

$$F(x) = \int_{-\infty}^{x} f(y)\, dy.$$

Substitute for $f(y)$ and write $u = (y-\mu)/\sigma$. We find that

$$F(x) = \int_{-\infty}^{x} \phi\{(y-\mu)/\sigma\}\, dy/\sigma = \int_{-\infty}^{t} \phi(u)\, du = \Phi(t).$$

This shows that $(x-\mu)/\sigma$ is dimensionless, and hence μ and σ both have the dimension of x. We also deduce that

$$F(b)-F(a) = \Phi\left(\frac{b-\mu}{\sigma}\right) - \Phi\left(\frac{a-\mu}{\sigma}\right).$$

Thus $p(a < X < b)$ can be calculated using a table of Φ.

Statistical methods are concerned not only with the original measurements but also with various functions of them which arise in practical applications. The normal distribution is important for three main reasons. First, it provides an adequate representation for many sets of data, especially under conditions where the frequency distribution of the measurements tails off symmetrically in each direction from the largest frequencies. Secondly, if the measurements are normally distributed, then the distributions of several functions of them arising in practice can be obtained exactly as a result of the mathematical properties of the normal distribution. Thirdly, whatever the distribution of the measurements, the distributions of many functions of them tend to normality when the number of measurements tends to infinity, and consequently such functions have distributions which are approximately normal for the number of measurements observed. The central place occupied by the normal distribution in statistics can be better appreciated when these brief descriptions of important aspects are supplemented by a fuller treatment in subsequent chapters.

Example 1. Measurements were made of the concentration in parts per million of copper in ores of the Altai mountains. For each concentration c, define

$$x = \log_2 c + 6{\cdot}5.$$

A frequency distribution of 688 values of x, taken from Vistelius

TABLE 6

Concentrations of copper in ores of the Altai mountains

Concentration in ppm (c)	x	*Observed frequency*	*Expected frequency*
0·016–	0·5–	2	3·8
0·031–	1·5–	25	19·3
0·062–	2·5–	65	66·6
0·125–	3·5–	127	142·5
0·250–	4·5–	198	189·2
0·500–	5·5–	168	156·0
1·000–	6·5–	76	79·7
2·000–	7·5–	24	25·2
4·000–	8·5–	2	5·0
8·000–16·000	9·5–10·5	1	0·6
Total		688	687·9

(1960), is given in the third column of Table 6. The expected frequencies in the fourth column are based on a normal distribution with $\mu = 5{\cdot}094$ and $\sigma = 1{\cdot}418$, for reasons which are explained in chapter 6. We illustrate the calculations involved by referring to the range of 0·250 to 0·500 ppm for c. This implies a range of 4·5 to 5·5 for x, and the corresponding values of $t = (x-\mu)/\sigma$ are $-0{\cdot}419$ and 0·286 respectively. Now $\Phi(0{\cdot}286) = 0{\cdot}6126$ and $\Phi(-0{\cdot}419) = 1-\Phi(0{\cdot}419) = 0{\cdot}3376$. Hence the expected frequency is

$$688(0{\cdot}6126-0{\cdot}3376) = 189{\cdot}2.$$

The two frequency distributions appear to be in good agreement.

Another method of examining whether the data agree with a normal distribution is as follows. Let $\Phi^{-1}(u)$ be the inverse function of $\Phi(t)$, defined by

$$\Phi^{-1}\{\Phi(t)\} = t.$$

$\Phi^{-1}(\varepsilon)$ is known as the **normal equivalent deviation** corresponding to the probability ε. The relationship $F(x) = \Phi\{(x-\mu)/\sigma\}$ implies that

$$\Phi^{-1}\{F(x)\} = (x-\mu)/\sigma.$$

For each x which determines the observed frequency distribution, $F(x)$ can be estimated by $\hat{F}(x)$, defined as the ratio of the cumulative frequency for measurements less than x to the total frequency. A plot of $y = \Phi^{-1}\{\hat{F}(x)\}$ against x gives a series of points which therefore lie close to a straight line. We do not require to estimate μ and σ. The procedure is simplified if the scale for y is labelled by standard probability levels, and this scale is commercially available on what is called normal probability paper.

TABLE 7

Normal equivalent deviations for copper data

x	*Cumulative frequency*	$\hat{F}(x)$	$\Phi^{-1}\{\hat{F}(x)\}$
1·5	2	0·00291	−2·76
2·5	27	0·0392	−1·76
3·5	92	0·134	−1·11
4·5	219	0·318	−0·47
5·5	417	0·606	0·27
6·5	585	0·850	1·04
7·5	661	0·9608	1·76
8·5	685	0·99564	2·62
9·5	687	0·99855	2·98

Example 2. The values of $\Phi^{-1}\{\hat{F}(x)\}$ for the data of example 1 are derived in Table 7. Fig. 9 shows that the corresponding points lie near a straight line fitted by inspection.

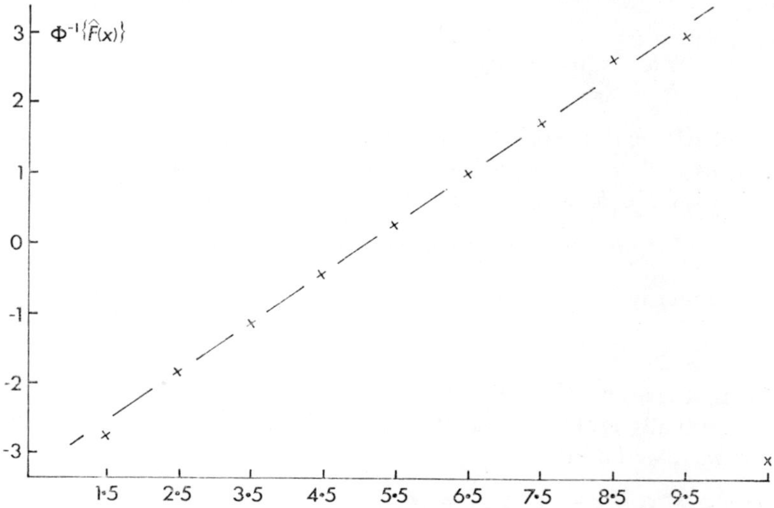

Fig. 9. Normal equivalent deviations for copper data.

§14. Several continuous variates

Consider an experiment where measurements made on a continuous scale appear in pairs, as for example the heights and weights of University students, or the daily maximum temperature and hours of sunlight at a weather station. We suppose that the outcome space for pairs of measurements is a plane. Define the variates X and Y as functions on the outcome space, and denote their distribution functions by $F(x)$ and $G(y)$ respectively. The probability that both $X \leqslant x$ and $Y \leqslant y$ is denoted by $H(x, y)$ and known as the **joint distribution** function of X and Y. Thus

$$p(X \leqslant x, Y \leqslant y) = H(x, y).$$

The event $(X \leqslant x, Y \leqslant y)$ tends to the event $X \leqslant x$ as $y \to \infty$ and to the event $Y \leqslant y$ as $x \to \infty$. Hence

$$\lim_{y \to \infty} H(x, y) = F(x) \quad \text{and} \quad \lim_{x \to \infty} H(x, y) = G(y).$$

From either of these results, we deduce that

$$\lim_{x, y \to \infty} H(x, y) = 1.$$

When $x \to -\infty$ or $y \to -\infty$, the event $(X \leqslant x, Y \leqslant y)$ becomes null. Hence

$$\lim_{x \to -\infty} H(x, y) = \lim_{y \to -\infty} H(x, y) = 0.$$

Let a, b, c, d be such that $a < b$ and $c < d$. The events $(X \leqslant a, Y \leqslant c)$ and $(a < X \leqslant b, Y \leqslant c)$ are mutually exclusive and their union is $(X \leqslant b, Y \leqslant c)$. Consequently

$$p(a < X \leqslant b, Y \leqslant c) = H(b, c) - H(a, c),$$

and similarly

$$p(a < X \leqslant b, Y \leqslant d) = H(b, d) - H(a, d).$$

Again, the events $(a < X \leqslant b, Y \leqslant c)$ and $(a < X \leqslant b, c < Y \leqslant d)$ are mutually exclusive, and their union is $(a < X \leqslant b, Y \leqslant d)$. In this way, we find that

$$p(a < X \leqslant b, c < Y \leqslant d) = H(b, d) - H(a, d) - H(b, c) + H(a, c).$$

Suppose that a continuous function $h(x, y)$ exists such that

$$H(x, y) = \int_{-\infty}^{x} \int_{-\infty}^{y} h(t, u)\, dt\, du.$$

The expression for $p(a < X \leqslant b, c < Y \leqslant d)$ can now be simplified, and when the constituent double integrals are combined, we obtain

$$p(a < X \leqslant b, c < Y \leqslant d) = \int_{a}^{b} dt \int_{c}^{d} h(t, u)\, dt\, du.$$

Fix b and let $a \to b$ from below. The limit of the double integral is zero, so that $p(X = b, c < Y \leqslant d) = 0$. Similarly

$$p(a < X \leqslant b, Y = d) = 0,$$

and therefore

$$p(a < X \leqslant b, c < Y \leqslant d) = p(a < X < b, c < Y < d).$$

A bivariate probability element is given by

$$p(x < X < x+dx, y < Y < y+dy) = h(x, y)\, dx\, dy.$$

The expression shows that $h(x, y) \geqslant 0$ for all x and y. Furthermore, the condition $H(\infty, \infty) = 1$ gives

$$\int_{-\infty}^{\infty} \int_{-\infty}^{\infty} h(x, y)\, dx\, dy = 1.$$

A function $h(x, y)$ with these properties defines **a continuous bivariate** distribution. We say that X and Y are continuous variates with the **joint density** function $h(x, y)$.

Denote the density functions of X and Y by $f(x)$ and $g(y)$ respectively. They can be derived from the joint density function of X and Y by using the results $F(x) = H(x, \infty)$ and $G(y) = H(\infty, y)$. Differentiation then gives

$$f(x) = \int_{-\infty}^{\infty} h(x, y)\, dy \quad \text{and} \quad g(y) = \int_{-\infty}^{\infty} h(x, y)\, dx,$$

which define the marginal distributions of X and Y respectively.

For any y such that $g(y) > 0$, we have

$$p(x < X < x+dx \mid y < Y < y+dy) = h(x, y)dx/g(y).$$

Denote $h(x, y)/g(y)$ by $f(x \mid y)$. This function defines a continuous probability distribution because $f(x \mid y) \geqslant 0$ and

$$\int_{-\infty}^{\infty} f(x \mid y)\, dx = \int_{-\infty}^{\infty} h(x, y)dx/g(y) = 1.$$

We say that $f(x \mid y)$ is the density function of X conditional on $Y = y$. The event $Y = y$ has zero probability, but the terminology is convenient. Similarly, for any x such that $f(x) > 0$, denote $h(x, y)/f(x)$ by $g(y \mid x)$. Then $g(y \mid x)$ is the density function of Y conditional on $X = x$.

The variates X and Y are described as **independent** if

$$h(x, y) = f(x)g(y) \quad \text{for all } x \text{ and } y.$$

Independence holds, for example, when X and Y are measurements made in independent experiments, because in this case

$$p(x < X < x+dx, y < Y < y+dy)$$
$$= p(x < X < x+dx)p(y < Y < y+dy).$$

If X and Y are independent variates, then

$$f(x \mid y) = f(x) \text{ for all } x \text{ and } y \text{ such that } g(y) > 0,$$

and

$$g(y \mid x) = g(y) \text{ for all } y \text{ and } x \text{ such that } f(x) > 0.$$

In general, we can define n variates $X_1, X_2, \ldots, X_n$ on the same outcome space. Their joint distribution function is

$$p(X_1 \leqslant x_1, X_2 \leqslant x_2, \ldots, X_n \leqslant x_n),$$

denoted by $H(x_1, x_2, \ldots, x_n)$. Suppose that a continuous function $h(x_1, x_2, \ldots, x_n)$ exists such that

$$H(x_1, x_2, \ldots, x_n)$$
$$= \int_{-\infty}^{x_1} \int_{-\infty}^{x_2} \ldots \int_{-\infty}^{x_n} h(u_1, u_2, \ldots, u_n)\, du_1\, du_2 \ldots du_n.$$

Then h is a non-negative function and its total integral is unity. We say that $h(x_1, x_2, \ldots, x_n)$ is the joint density function of the continuous variates $X_1, X_2, \ldots, X_n$.

The **marginal density** function for any set of variates is obtained from $h(x_1, x_2, \ldots, x_n)$ by integrating out the superfluous variables. In particular, the density function of X_i is

$$f_i(x_i) = \int_{-\infty}^{\infty} \int_{-\infty}^{\infty} \ldots \int_{-\infty}^{\infty} h(x_1, x_2, \ldots, x_n) \prod_{j \neq i} dx_j.$$

The variates $X_1, X_2, \ldots, X_n$ are said to be **mutually independent** if

$$h(x_1, x_2, \ldots, x_n) = f_1(x_1) f_2(x_2) \ldots f_n(x_n)$$

for all $x_1, x_2, \ldots, x_n$.

§15. Normal bivariate distribution

Consider the identity

$$t^2 - 2\rho tu + u^2 \equiv t^2(1-\rho^2) + (u - \rho t)^2.$$

Suppose that ρ is fixed and such that $-1 < \rho < 1$. Then $t^2 - 2\rho tu + u^2$ is non-negative for all t and u. Its minimum value, namely zero, is attained only if $t = u = 0$. Now define $\psi(t, u)$ by

$$\psi(t, u) = (2\pi)^{-1}(1-\rho^2)^{-\frac{1}{2}} \exp\{-(t^2 - 2\rho tu + u^2)/2(1-\rho^2)\}.$$

Then ψ is a positive function which attains its maximum when $t = u = 0$. Using the identity, we find that

$$\int_{-\infty}^{\infty} \psi(t, u)\, du = (2\pi)^{-1}(1-\rho^2)^{-\frac{1}{2}} \exp(-\tfrac{1}{2}t^2)$$
$$\times \int_{-\infty}^{\infty} \exp\{-(u-\rho t)^2/2(1-\rho^2)\}\, du.$$

Substitute $w = (u-\rho t)(1-\rho^2)^{-\frac{1}{2}}$ in the integral on the right, which becomes

$$(1-\rho^2)^{\frac{1}{2}} \int_{-\infty}^{\infty} \exp(-\tfrac{1}{2}w^2)\, dw = \{2\pi(1-\rho^2)\}^{\frac{1}{2}}.$$

We therefore obtain

$$\int_{-\infty}^{\infty} \psi(t, u)\, du = (2\pi)^{-\frac{1}{2}} \exp(-\tfrac{1}{2}t^2),$$

and hence

$$\int_{-\infty}^{\infty} \int_{-\infty}^{\infty} \psi(t, u)\, dt\, du = 1.$$

Thus ψ has the properties required of a function which defines a continuous bivariate distribution.

The variates T and U with joint density function ψ are said to have a **standard normal bivariate** distribution. Both T and U individually have standard normal distributions because

$$\int_{-\infty}^{\infty} \psi(t, u)\, du = \phi(t)$$

and there is a similar result when ψ is integrated over t. The density function of U conditional on $T = t$ is

$$\psi(t, u)/\phi(t) = \{2\pi(1-\rho^2)\}^{-\frac{1}{2}} \exp\{-(u-\rho t)^2/2(1-\rho^2)\}.$$

This is the density function of a normal distribution with parameters $\mu = \rho t$ and $\sigma^2 = (1-\rho^2)$. Thus T and U are independent variates if and only if $\rho = 0$.

We have proved that

$$\int_{-\infty}^{\infty} \int_{-\infty}^{\infty} (2\pi)^{-1} (1-\rho^2)^{-\frac{1}{2}} \exp\{-(t^2-2\rho\, tu+u^2)/2(1-\rho^2)\}\, dt\, du = 1.$$

Substitute $t = (x-\mu_x)/\sigma_x$ and $u = (y-\mu_y)/\sigma_y$, where $-\infty < \mu_x < \infty$, $-\infty < \mu_y < \infty$, $\sigma_x > 0$, and $\sigma_y > 0$. The total integral then takes the form

$$\int_{-\infty}^{\infty} \int_{-\infty}^{\infty} h(x, y)\, dx\, dy = 1,$$

where $h(x, y)$ denotes the function

$$(2\pi\sigma_x\sigma_y)^{-1}(1-\rho^2)^{-\frac{1}{2}}$$
$$\times \exp\left[-\left\{\left(\frac{x-\mu_x}{\sigma_x}\right)^2 - 2\rho\left(\frac{x-\mu_x}{\sigma_x}\right)\left(\frac{y-\mu_y}{\sigma_y}\right) + \left(\frac{y-\mu_y}{\sigma_y}\right)^2\right\}/2(1-\rho^2)\right]$$

Therefore $h(x, y)$ defines a continuous bivariate distribution.

We say that the variates X and Y with joint density function $h(x, y)$ have a **normal bivariate** distribution with parameters μ_x, μ_y, σ_x^2, σ_y^2, and ρ. When $\mu_x = \mu_y = 0$ and $\sigma_x = \sigma_y = 1$, X and Y have a standard normal bivariate distribution. The functions $h(x, y)$ and $\psi(t, u)$ are related by

$$h(x, y)\, dx\, dy = \psi(t, u)\, dt\, du.$$

When x is fixed, so is t, and when y increases, so does u. Consequently

$$dx \int_{-\infty}^{\infty} h(x, y)\, dy = dt \int_{-\infty}^{\infty} \psi(t, u)\, du.$$

We have seen that the integral on the right is $\phi(t)$, whence

$$f(x)\, dx = \phi(t)\, dt.$$

This shows that

$$f(x) = (2\pi\sigma_x^2)^{-\frac{1}{2}} \exp\{-(x-\mu_x)/2\sigma_x^2\}.$$

Thus X is normally distributed with parameters μ_x and σ_x^2. Similarly Y is normally distributed with parameters μ_y and σ_y^2.

We now determine $g(y \mid x)$, the density function of Y conditional on $X = x$. From the definition,

$$g(y \mid x)\, dy = h(x, y)\, dx\, dy/f(x)\, dx.$$

The ratio on the right is $\psi(t, u)\, dt\, du/\phi(t)\, dt$. When the expression for $\psi(t, u)/\phi(t)$ is substituted, we obtain

$$g(y \mid x)\, dy = \{2\pi(1-\rho^2)\}^{-\frac{1}{2}} \exp\{-(u-\rho t)^2/2(1-\rho^2)\}\, du.$$

Therefore

$$g(y \mid x) = \{2\pi\sigma_y^2(1-\rho^2)\}^{-\frac{1}{2}} \exp\left[-\left\{\left(\frac{y-\mu_y}{\sigma_y}\right) - \rho\left(\frac{x-\mu_x}{\sigma_x}\right)\right\}^2/2(1-\rho^2)\right].$$

This is the density function of a normal distribution with parameters

$$\mu = \mu_y + \rho\sigma_y(x-\mu_x)/\sigma_x \quad \text{and} \quad \sigma^2 = \sigma_y^2(1-\rho^2).$$

EXERCISES 3

1. Measurements were made of the intervals between α-particle emissions from americium-241. The frequency distribution in Table 8 is a condensed version of one given by Berkson (1966), and

TABLE 8

Intervals between emission of α-particles by americium-241

Time (sec.)	*Frequency*	*Time (sec.)*	*Frequency*
0–0·1	5803	0·5–0·6	102
0·1–0·2	2519	0·6–0·7	28
0·2–0·3	1071	0·7–0·8	16
0·3–0·4	457	0·8–∞	16
0·4–0·5	208		
		Total	10220

an estimate of λ is 8·398 sec.$^{-1}$. Calculate the expected frequencies and compare them with the observed frequencies.

2. Plot $y = \Phi^{-1}\{\hat{F}(x)\}$ against x using the values in Table 7. Fit a straight line by inspection and estimate μ and σ from the slope of the line and the intercept on the y-axis. Compare the estimates with those given in §**13**.

3. The weight of a manufactured item is normally distributed with parameters μ and σ, where σ is fixed but μ is adjustable. For a given value of μ, the average cost of manufacture is $a+b\mu$. An item which exceeds w in weight is sold, but if the weight is less than w then the item is discarded at a cost of c. Show that the average cost of an item sold is

$$k = \left\{a+b\mu+c\Phi\left(\frac{w-\mu}{\sigma}\right)\right\}\Big/\left\{1-\Phi\left(\frac{w-\mu}{\sigma}\right)\right\}.$$

Deduce that the value of μ which minimizes k is $w-x\sigma$, where x is the negative root of the equation

$$x+\{1-\Phi(x)\}/\phi(x) = \{w+(a+c)/b\}/\sigma.$$

4. Suppose that incidents occur randomly at instants of time under the conditions described in §**12**. Assume also that the probability for two or more incidents to occur in an interval of duration h is $o(h)$ as $h \to 0$. Here $o(h)$ is such that $o(h)/h \to 0$ as $h \to 0$. Let $p_n(x)$ be the probability that n incidents occur in an interval of duration x. Show that

$$p_1(h) = \lambda h + o(h) \quad \text{as} \quad h \to 0.$$

5. The assumptions and notation are the same as in the previous exercise. Let $f_n(x)$ be the density function of the interval which begins just after an incident occurs and ends just after the nth subsequent incident occurs. Show that $f_n(x) = \lambda p_{n-1}(x)$ and

$$p_n(x) = \int_0^x f_n(s)\, e^{-\lambda(x-s)}\, ds.$$

Starting with $f_1(x) = \lambda e^{-\lambda x}$, prove by induction that

$$f_n(x) = \lambda^n x^{n-1} e^{-\lambda x}/(n-1)!.$$

6. A **mixed** bivariate distribution arises when one variate is continuous and the other is discrete. Suppose that X has the density function

$$f(x) = \begin{cases} \lambda^n x^{n-1} e^{-\lambda x}/(n-1)! & (x > 0;\ \lambda > 0) \\ 0 & (x \leqslant 0) \end{cases}$$

where n is a positive integer. Let the distribution of Y conditional on $X = x$ be Poisson with parameter x. Show that the unconditional distribution of Y is a negative binomial given by

$$p(Y = k) = \binom{n+k-1}{n-1}\left(\frac{\lambda}{\lambda+1}\right)^n\left(\frac{1}{\lambda+1}\right)^k \quad (k = 0, 1, 2, \ldots).$$

Find the distribution of X conditional on $Y = k$.

7. The continuous variates X and Y have the joint density function $h(x, y)$. Show that $h(x, y)$ is given by any of the following expressions:

$$\frac{\partial^2}{\partial x\, \partial y}\, p(X < x,\ Y < y), \qquad \frac{\partial^2}{\partial x\, \partial y}\, p(X > x,\ Y > y),$$

$$-\frac{\partial^2}{\partial x\, \partial y}\, p(X > x,\ Y < y), \qquad -\frac{\partial^2}{\partial x\, \partial y}\, p(X < x,\ Y > y).$$

8. The density function of a standard normal bivariate distribution is $\psi(t, u)$. Prove that

$$\frac{\partial \psi}{\partial \rho} = \frac{\partial^2 \psi}{\partial t\, du}.$$

Use this result to show that

$$\int_{-\infty}^{0}\int_{-\infty}^{0} \psi(t, u)\, dt\, du = \{\cos^{-1}(-\rho)\}/2\pi.$$

9. A bivariate distribution is **singular** when there is a one-one correspondence between the variates. Suppose that X and Y have a standard normal bivariate distribution, and write

$$\Psi(a, b) = \int_{-\infty}^{a} \int_{-\infty}^{b} \psi(t, u)\, dt\, du.$$

Show that $X = Y$ in the limit as $\rho \to 1$, and $X = -Y$ in the limit as $\rho \to -1$. Deduce that

$$\lim_{\rho \to 1} \Psi(a, b) = \min\{\Phi(a), \Phi(b)\},$$

and

$$\lim_{\rho \to -1} \Psi(a, b) = \max\{0, \Phi(a)+\Phi(b)-1\}.$$

10. Given the distribution functions $F(x)$ and $G(y)$, suppose that $H(x, y)$ satisfies the equation

$$\delta = \frac{H(1-F-G+H)}{(F-H)(G-H)},$$

where δ has any fixed value in $(0, \infty)$. Prove that there is just one solution $H(x, y)$ which is a valid joint distribution function and has the marginal distribution functions $F(x)$ and $G(y)$.

CHAPTER 4
PROPERTIES OF VARIATES

§16. Mean values

The concept of a mean value is familiar in everyday life, and the calculation of means has already appeared incidentally when estimating the parameters of probability distributions. Thus, when comparing the frequency distribution of the number of boys among the first 5 children in families of size 5 or more with a binomial probability distribution for which $n = 5$, the parameter θ was estimated by the observed proportion of boys in 2571 families. This is equivalent to estimating $n\theta$ by the average number of boys per family, because the average number is n times the proportion of boys. We now give explicit definitions of mean value for both finite sets of measurements and probability distributions.

Suppose first that r observations are made on a variate X with the probability distribution

$$p(X = j) = f_j \qquad (j = 0, 1, 2, \ldots).$$

Let the measurements consist of 0 with frequency r_0, 1 with frequency r_1, 2 with frequency $r_2, \ldots$, where

$$r_0 + r_1 + r_2 + \ldots = r.$$

The set of r measurements forms a **sample** of **size** r from the distribution of X. We define the **sample mean** m by

$$m = (r_1 + 2r_2 + 3r_3 + \ldots)/r.$$

The sample mean is an important quantity in statistics and its properties are subsequently examined in detail. For the present we are concerned with the form m takes as $r \to \infty$. The frequency-ratios

$r_1/r, r_2/r, \ldots$ then tend to the probabilities $f_1, f_2, \ldots$ respectively and the corresponding limit for m is $\Sigma j f_j$. This argument suggests the introduction of a quantity $E(X)$ which is called the **expectation** of X and given by

$$E(X) = \sum_{j=1}^{\infty} j f_j.$$

In general, for a discrete variate X with the distribution

$$p(X = x_j) = f_j \qquad (j = 0, 1, 2, \ldots),$$

we define the expectation of X by

$$E(X) = \sum_{j=0}^{\infty} x_j f_j,$$

provided that the series is absolutely convergent. If the series is not absolutely convergent, then $E(X)$ is undefined. Other terms for $E(X)$ are the **mean** of X, the **expected value** of X, and the **distribution mean**.

Expectation is a term in standard usage which is as old as the theory of probability. That theory first originated in games of chance played for money, and expectation refers to the mean gain or loss of the gambler. Actuarial science began at about the same time using similar ideas, and the expectation of life at age t is simply the mean number of years lived thereafter by people who survive to age t.

Example 1. X has the binomial distribution

$$p(X = j) = \binom{n}{j} \theta^j (1-\theta)^{n-j} \qquad (j = 0, 1, \ldots, n).$$

There are zero probabilities for $j > n$, whence

$$E(X) = \sum_{j=1}^{n} j \binom{n}{j} \theta^j (1-\theta)^{n-j}.$$

Make the substitution

$$j\binom{n}{j} = \frac{j\,n!}{(n-j)!\,j!} = \frac{n(n-1)!}{(n-j)!\,(j-1)!} = n\binom{n-1}{j-1}.$$

We obtain

$$\begin{aligned} E(X) &= n\theta \sum_{j=1}^{n} \binom{n-1}{j-1} \theta^{j-1} (1-\theta)^{(n-1)-(j-1)} \\ &= n\theta\{\theta + (1-\theta)\}^{n-1} = n\theta. \end{aligned}$$

This result supports the estimation of $n\theta$ by the sample mean, which was the method used in §**6**.

Suppose next that r observations are made on a continuous variate X with density function $f(x)$. Denote the resulting measurements by $x_1, x_2, \ldots, x_r$. They constitute a sample of size r from the distribution of X, and the sample mean m is defined by

$$m = (x_1+x_2+\ldots+x_r)/r.$$

We are interested again in the form that m tends towards as $r \to \infty$. Since the probability that X takes any specified value is zero, we divide the range of variation of X into equal small intervals of length h, and then replace the distribution of X by a discrete probability distribution where a typical interval $(x, x+h)$ is associated with probability $hf(x)$. This is an approximate representation which improves as $h \to 0$. The frequency-ratio of observations on X which lie in $(x, x+h)$ tends to $hf(x)$ as $r \to \infty$, and the limit implied for m is $\Sigma xhf(x)$. Taking a further limit as $h \to 0$, we obtain $\int xf(x)\,dx$. Although the argument is far from rigorous, it suffices to motivate a definition of the expectation of X for a continuous probability distribution, namely

$$E(X) = \int_{-\infty}^{\infty} xf(x)\,dx.$$

The expectation of X is finite if the integral converges absolutely, and otherwise $E(X)$ is undefined.

Example 2. X has the exponential distribution with density function

$$f(x) = \begin{cases} \lambda e^{-\lambda x} & (x > 0), \\ 0 & (x \leqslant 0). \end{cases}$$

We omit from the range of integration for $E(X)$ any interval where $f(x) = 0$. Hence

$$E(X) = \int_0^{\infty} x\lambda e^{-\lambda x}\,dx.$$

Make the substitution $\lambda x = u$, whence

$$E(X) = \lambda^{-1}\int_0^{\infty} ue^{-u}\,du$$

$$= \lambda^{-1}[-(u+1)e^{-u}]_0^{\infty} = \lambda^{-1}.$$

This result suggests the estimation of λ by the reciprocal of the sample mean. However, a different method is used to estimate λ in §**12**, where the data consist of a frequency distribution and the original measurements are not available. The method is explained in chapter 11.

§17. Moments for one variate

We are concerned with the expectations of powers of X in this section, and the expectations of other functions such as log X are also required in practice. Suppose that $t(x)$ is a specified function of x, and X is a continuous variate with density function $f(x)$. The existence of a continuous variate $T = t(X)$ is established in §**21** by finding its density function $c(t)$. Thus

$$E(T) = \int_{-\infty}^{\infty} tc(t)\, dt.$$

However, $E(T)$ can be obtained directly from the distribution of X by the expression

$$E(T) = \int_{-\infty}^{\infty} t(x)f(x)\, dx,$$

which can be motivated by an argument similar to that which preceded the definition of $E(X)$. This formula agrees with the result from the distribution of T if

$$\int_{-\infty}^{\infty} tc(t)\, dt = \int_{-\infty}^{\infty} t(x)f(x)\, dx,$$

which is proved in §**21** under general conditions.

Similarly, when X is a discrete variate with distribution

$$p(X = x_j) = f_j \qquad (j = 0, 1, 2, \ldots),$$

then $T = t(X)$ is a discrete variate with distribution

$$p(T = t_k) = c_k \qquad (k = 0, 1, 2, \ldots).$$

The expectation of T is

$$E(T) = \sum_{k=0}^{\infty} t_k c_k,$$

but $E(T)$ can be found more easily from the alternative expression

$$E(T) = \sum_{j=0}^{\infty} t(x_j)f_j.$$

We have assumed that any variates considered have finite expectations, and we continue to make this assumption in what follows except where the contrary is specifically noted. Results about expectation will be proved for continuous variates only. The convention that expectations are finite permits the rearrangement of multiple integrals without further comment. All the corresponding proofs for discrete variates are straightforward analogies in which probability elements are replaced by probabilities, integrations by summations, and the rearrangement of multiple sums is justified.

For example, suppose that functions $u_1, u_2, \ldots$ exist such that

$$t(x) = \Sigma\, a_i u_i(x) + b,$$

where $\{a_i\}$ and b are constants. Then

$$E\{t(X)\} = \Sigma\, a_i E\{u_i(X)\} + b,$$

because

$$\int \{\Sigma\, a_i u_i(x) + b\} f(x)\, dx = \Sigma\, a_i \{ \int u_i(x) f(x)\, dx\} + b.$$

In particular, we have

$$E(aX+b) = aE(X)+b.$$

The rth **moment** of X **about zero** is denoted by μ'_r and defined by

$$\mu'_r = E(X^r) \qquad (r = 0, 1, 2, \ldots).$$

Thus $\mu'_0 = 1$ and μ'_1 is the mean of X. Write $\mu'_1 = \mu$. The rth **central moment** of X is denoted by μ_r and defined by

$$\mu_r = E\{(X-\mu)^r\} \qquad (r = 0, 1, 2, \ldots).$$

Thus $\mu_0 = 1$ and

$$\mu_1 = E(X-\mu) = E(X)-\mu = 0.$$

The second central moment is called the **variance** of X and written as var (X). Thus

$$\text{var}\,(X) = E(X-\mu)^2.$$

An alternative expression is

$$\text{var}\,(X) = E(X^2)-\mu^2,$$

because the formula for $E\{\Sigma\, a_i u_i(X)+b\}$ implies that

$$E(X^2-2\mu X+\mu^2) = E(X^2)-2\mu\, E(X)+\mu^2 = E(X^2)-\mu^2.$$

Since var (X) is the expectation of a non-negative variate $(X-\mu)^2$, it must also be non-negative. The non-negative square root of var (X) is called the **standard deviation** of X and is measured in the same units as X. A common shorthand is var $(X) = \sigma^2$, in which case the standard deviation is σ. The result

$$\text{var}\,(aX+b) = a^2\,\text{var}\,(X)$$

is often used. It is proved by noting that $E(aX+b) = a\mu+b$, whence

$$\text{var}\,(aX+b) = E(aX+b-a\mu-b)^2 = a^2E(X-\mu)^2 = a^2\,\text{var}\,(X).$$

The standard deviation of $aX+b$ is $|\,a\,|$ times that of X.

When the values taken by X are non-negative integers, a system of **factorial moments** is convenient. The rth factorial moment of X is denoted by $\mu_{(r)}$ and defined by

$$\mu_{(r)} = E\{X(X-1)(X-2)\ldots(X-r+1)\} \qquad (r = 1, 2, \ldots).$$

Another expression for the variance of X is thus

$$\text{var}\,(X) = \mu_{(2)}+\mu-\mu^2,$$

because the right-hand side is

$$E(X^2)-E(X)+\mu-\mu^2 = E(X^2)-\mu^2.$$

Example 1. X has the binomial distribution

$$p(X=j) = \binom{n}{j}\theta^j(1-\theta)^{n-j} \qquad (j = 0, 1, \ldots, n).$$

The second factorial moment is

$$\mu_{(2)} = \sum_{j=2}^{n} j(j-1)\binom{n}{j}\theta^j(1-\theta)^{n-j}.$$

Proceeding as before, we obtain

$$\mu_{(2)} = n(n-1)\theta^2\sum_{j=2}^{n}\binom{n-2}{j-2}\theta^{j-2}(1-\theta)^{(n-2)-(j-2)}$$

$$= n(n-1)\theta^2\{\theta+(1-\theta)\}^{n-2} = n(n-1)\theta^2.$$

Hence

$$\text{var}\,(X) = n(n-1)\theta^2+n\theta-(n\theta)^2 = n\theta(1-\theta).$$

When $\theta = \lambda/n$ and $n \to \infty$, X has the Poisson distribution

$$p(X = j) = \lambda^j e^{-\lambda}/j! \qquad (j = 0, 1, 2, \ldots).$$

The same limiting process gives $n\theta = \lambda$ and $n\theta(1-\theta) \to \lambda$. We therefore find that the mean and variance of a Poisson distribution are both equal to the parameter λ.

Example 2. X has the normal distribution with density function

$$f(x) = (2\pi\sigma_1^2)^{-\frac{1}{2}} \exp\{-(x-\mu_1)^2/2\sigma_1^2\} \quad (-\infty < x < \infty).$$

We first evaluate

$$E[\{(X-\mu_1)/\sigma_1\}^r] = \int_{-\infty}^{\infty} \{(x-\mu_1)/\sigma_1\}^r f(x)\, dx$$

when $r = 1$ and 2. Substitute $t = (x-\mu_1)/\sigma_1$, so that

$$E[\{(X-\mu_1)/\sigma_1\}^r] = \int_{-\infty}^{\infty} t^r \phi(t)\, dt,$$

where $\phi(t)$ is the standard normal density function. Now

$$\frac{d\phi}{dt} = \frac{d}{dt}\{(2\pi)^{-\frac{1}{2}} e^{-\frac{1}{2}t^2}\} = -t\phi(t).$$

Hence the integral for $r = 1$ is

$$\int_{-\infty}^{\infty} t\phi(t)\, dt = [-\phi(t)]_{-\infty}^{\infty} = 0,$$

and integration by parts when $r = 2$ gives

$$\int_{-\infty}^{\infty} t\left(-\frac{d\phi}{dt}\right) dt = [-t\phi(t)]_{-\infty}^{\infty} + \int_{-\infty}^{\infty} \phi(t)\, dt = 1.$$

These results show that

$$E\{(X-\mu_1)/\sigma_1\} = 0 \quad \text{and} \quad E\{(X-\mu_1)^2/\sigma_1^2\} = 1.$$

The first implies that $E(X) = \mu_1$ and the second then gives var $(X) = \sigma_1^2$. Our previous notation for the normal density function was

$$f(x) = (2\pi\sigma^2)^{-\frac{1}{2}} \exp\{-(x-\mu)^2/2\sigma^2\}.$$

This notation is therefore consistent with writing μ for mean and σ^2 for variance.

We have seen that $E\{t(X)\}$ is determined when the distribution of X is known. In some cases, the integral or sum which is involved cannot be evaluated exactly and other methods of approach must be considered. The calculations can always be done numerically for specified $t(x)$ and $f(x)$, but a simple and general approximation is convenient. We proceed to derive an approximate formula for $E\{t(X)\}$ in terms of the mean and variance of X. This has the additional advantage that only these properties of X are necessary for an evaluation of $E\{t(X)\}$. The first step in deriving the formula consists of expanding $t(x)$ in a Taylor series about the point μ. Thus

$$t(x) = t(\mu)+(x-\mu)t'(\mu)+\tfrac{1}{2}(x-\mu)^2t''(\mu)+\tfrac{1}{6}(x-\mu)^3t'''(z),$$

where the primes signify differentiation and z is a value between x and μ. Now multiply both sides by $f(x)$ and integrate over x. Since $E(X-\mu) = 0$, the result is

$$E\{t(X)\} = t(\mu)+\tfrac{1}{2}\sigma^2t''(\mu)+\tfrac{1}{6}\mu_3t'''(\xi).$$

Here ξ is an unknown value of x, and the approximation is obtained by ignoring the last term on the right. Hence we obtain

$$E\{t(X)\} \simeq t(\mu)+\tfrac{1}{2}\sigma^2t''(\mu).$$

This formula is likely to be satisfactory provided that the linear and quadratic terms in the Taylor series adequately account for the variation of $t(x)$ with x.

Example 3. X is normally distributed and $t(X) = (X-\mu)^2$. We know that $E\{t(X)\} = \sigma^2$ exactly. The formula gives

$$E\{t(X)\} \simeq \tfrac{1}{2}\sigma^2(2) = \sigma^2,$$

which agrees with the exact result. On the other hand, the function $t(X) = (X-\mu)^4$ does not satisfy the conditions under which the approximation is likely to be satisfactory. The formula gives $E\{t(X)\} \simeq 0$, whereas direct integration leads to $E\{t(X)\} = 3\sigma^4$. A proof of the correct result appears in §**25**.

The same method can be used to derive an approximation for var $\{t(X)\}$. We suppose that $t''(\mu)$ is small, whence

$$t(x) \simeq t(\mu)+(x-\mu)t'(\mu) \quad \text{and} \quad E\{t(X)\} \simeq t(\mu).$$

Under these conditions,

$$\begin{aligned}\text{var}\,\{t(X)\} &= E[t(X)-E\{t(X)\}]^2\\ &\simeq \{t'(\mu)\}^2E(X-\mu)^2 = \{t'(\mu)\}^2\,\text{var}\,(X).\end{aligned}$$

The approximation is thus

$$\operatorname{var}\{t(X)\} \simeq \{t'(\mu)\}^2 \operatorname{var}(X).$$

Both sides are equal when $t(X) = aX+b$.

§18. Measures of location and spread

The mean and variance of a probability distribution are important for the following reasons. First, as already remarked in §**13**, the distributions of many functions of the measurements are approximately normal when the number of measurements is large. Provided that the mean and variance of any such function can be obtained, the entire distribution is known. Secondly, the mean and variance have useful descriptive aspects when we want to summarize the properties of a distribution in a few numerical statements. When we need a single representative value for X, the mean is an appropriate quantity. Again, when we need a single value for the amount by which the distribution of X is dispersed over the possible values of X, the variance or standard deviation is appropriate because the magnitude of $E(X-\mu)^2$ will depend on whether most of the distribution is near the mean or whether large values of $|X-\mu|$ are likely. We say that the mean is a **measure of location**, and quantities such as the variance and standard deviation are called **measures of dispersion** or **spread**. Other measures of location and spread are defined as follows.

Suppose that X is a continuous variate with distribution function $F(x)$, and let ε be any number in (0, 1). We assume that the equation

$$F(\xi_\varepsilon) = 1-\varepsilon$$

has a unique solution ξ_ε. This condition is fulfilled for most of the continuous distributions which arise in statistical practice. The solution ξ_ε is known as the **upper** 100ε **percent point** of the distribution of X, because values larger than ξ_ε occur with total probability ε. Similarly, $\xi_{1-\varepsilon}$ is the **lower** 100ε **percent point**. The upper and lower 50 percent points coincide, and we say that $\xi_{0\cdot5}$ is the **distribution median.** This quantity is representative of X and the median therefore qualifies as a measure of location. If $\varepsilon_1 < \varepsilon_2$, then $\xi_{\varepsilon_1} > \xi_{\varepsilon_2}$. The difference $\xi_{\varepsilon_1}-\xi_{\varepsilon_2}$ is thus another measure of spread.

A distribution is described as **symmetrical** if

$$F(x) = 1-F(2\xi_{0\cdot5}-x) \quad \text{for all } x,$$

where we note that the points x and $2\xi_{0\cdot 5}-x$ are equidistant from $\xi_{0\cdot 5}$. For example, any normal distribution is symmetrical. If $\xi_{0\cdot 5} = 0$, then

$$F(x) = 1-F(-x) \quad \text{for all } x.$$

Take $\varepsilon < \frac{1}{2}$ and substitute $x = \xi_\varepsilon$, whence $F(-\xi_\varepsilon) = \varepsilon$. We therefore find that $\xi_{1-\varepsilon} = -\xi_\varepsilon$, and

$$p(-\xi_\varepsilon < X < \xi_\varepsilon) = 1-2\varepsilon.$$

§19. Moments for several variates

We begin with some general remarks which extend those concerning $E\{t(X)\}$ made at the beginning of §**17**. Suppose that $t(x, y)$ is a specified function of x and y. When X and Y are continuous variates with the joint density function $h(x, y)$, the existence of a continuous variate $T = t(X, Y)$ is established by finding its density function, and the results imply that

$$E\{t(X, Y)\} = \int_{-\infty}^{\infty} \int_{-\infty}^{\infty} t(x, y)\, h(x, y)\, dx\, dy.$$

The problem of finding the distribution of $t(X, Y)$ is discussed in §**22**, and we prove then that this expression for $E\{t(X, Y)\}$ is consistent with the one obtained directly from the distribution of T. Similarly, when X and Y are discrete variates with the joint distribution

$$p(X = x_j, Y = y_k) = h_{jk},$$

we obtain

$$E\{t(X, Y)\} = \sum_{j=0}^{\infty} \sum_{k=0}^{\infty} t(x_j, y_k) h_{jk}.$$

Two special cases can usefully be noted. First, consider $E\{u(X)+v(Y)\}$, which is given by

$$\begin{aligned}\iint \{u(x)+v(y)\}\, h(x, y)\, dx\, dy \\ &= \int u(x)\, dx \int h(x, y)\, dy + \int v(y)\, dy \int h(x, y)\, dx \\ &= \int u(x) f(x)\, dx + \int v(y) g(y)\, dy.\end{aligned}$$

We thus find that

$$E\{u(X)+v(Y)\} = E\{u(X)\}+E(v(Y)\}.$$

Secondly, suppose that X and Y are independent variates. Consider $E\{u(X)\,v(Y)\}$, which is given by

$$\begin{aligned}\iint u(x)v(y)h(x, y)\,dx\,dy &= \iint u(x)v(y)f(x)g(y)\,dx\,dy \\ &= \{\int u(x)f(x)\,dx\}\{\int v(y)g(y)\,dy\}.\end{aligned}$$

The independence of X and Y therefore implies that

$$E\{u(X)v(Y)\} = E\{u(X)\}E\{v(Y)\}.$$

Denote the means of X and Y by μ_x and μ_y respectively. The concept of moments is extended to two variates by considering quantities of the form

$$E(X^r Y^s) \quad \text{or} \quad E\{(X-\mu_x)^r(Y-\mu_y)^s\}.$$

In practice only the case $r = s = 1$ is important. The **covariance** of X and Y is written cov (X, Y) and defined by

$$\text{cov}\,(X, Y) = E\{(X-\mu_x)(Y-\mu_y)\}.$$

An alternative expression is

$$\text{cov}\,(X, Y) = E(XY)-\mu_x\mu_y,$$

because the expectation of $XY-\mu_x Y-\mu_y X+\mu_x\mu_y$ is

$$E(XY)-\mu_x E(Y)-\mu_y E(X)+\mu_x\mu_y = E(XY)-\mu_x\mu_y.$$

In the same way, we obtain

$$\text{cov}\,(aX+b, cY+d) = ac\,\text{cov}\,(X, Y).$$

If X and Y are independent, then $E(XY) = E(X)E(Y)$, so that cov (X, Y) is zero.

Example 1. T and U have the standard normal bivariate distribution with density function

$$\psi(t, u) = (2\pi)^{-1}(1-\rho^2)^{-\frac{1}{2}} \exp\{-(t^2-2\rho tu+u^2)/2(1-\rho^2)\}.$$

Some properties of this distribution have already been derived in §**15**. In particular, T and U each have the density function ϕ corresponding to a normal distribution with parameters $\mu = 0$ and $\sigma^2 = 1$. The results of example 2 in §**17** show that

$$E(T) = E(U) = 0 \quad \text{and} \quad \text{var}\,(T) = \text{var}\,(U) = 1.$$

Now consider $E(TU)$. We use the factorization

$$\psi(t, u) = \phi(t)\{2\pi(1-\rho^2)\}^{-\frac{1}{2}} \exp\{-(u-\rho t)^2/2(1-\rho^2)\},$$

where the conditional density function $\psi(t, u)/\phi(t)$ is that of a normal distribution with parameters $\mu = \rho t$ and $\sigma^2 = (1-\rho^2)$. Hence

$$E(TU) = \int_{-\infty}^{\infty}\int_{-\infty}^{\infty} tu\psi(t, u)\,dt\,du$$

$$= \int_{-\infty}^{\infty} t\phi(t)\,dt[\int_{-\infty}^{\infty} u\{2\pi(1-\rho^2)\}^{-\frac{1}{2}} \exp\{-(u-\rho t)^2/2(1-\rho^2)\}\,du].$$

The value of the inner integral is ρt and therefore

$$E(TU) = \rho \int_{-\infty}^{\infty} t^2\phi(t)\,dt = \rho.$$

We conclude that cov $(T, U) = \rho$ also.

Example 2. X and Y have a normal bivariate distribution with density function $h(x, y)$ given by

$$(2\pi\sigma_x\sigma_y)^{-1}(1-\rho^2)^{-\frac{1}{2}}$$

$$\exp\left[-\left\{\left(\frac{x-\mu_x}{\sigma_x}\right)^2 - 2\rho\left(\frac{x-\mu_x}{\sigma_x}\right)\left(\frac{y-\mu_y}{\sigma_y}\right)+\left(\frac{y-\mu_y}{\sigma_y}\right)^2\right\}\Big/2(1-\rho^2)\right].$$

We showed in §**15** that X has a normal distribution with parameters μ_x and σ_x^2, and hence $E(X) = \mu_x$, var $(X) = \sigma_x^2$. Similarly $E(Y) = \mu_y$ and var $(Y) = \sigma_y^2$. Now consider

$$E\left\{\left(\frac{X-\mu_x}{\sigma_x}\right)\left(\frac{Y-\mu_y}{\sigma_y}\right)\right\} = \int_{-\infty}^{\infty}\int_{-\infty}^{\infty}\left(\frac{x-\mu_x}{\sigma_x}\right)\left(\frac{y-\mu_y}{\sigma_y}\right) h(x, y)\,dxdy.$$

Making the same substitutions as before, we obtain

$$E\{(X-\mu_x)(Y-\mu_y)\} = \sigma_x\sigma_y \int_{-\infty}^{\infty}\int_{-\infty}^{\infty} tu\psi(t, u)\,dt\,du.$$

According to example 1, the value of the double integral is ρ. Thus cov $(X, Y) = \rho\sigma_x\sigma_y$. We see that all the parameters of a normal bivariate distribution can be expressed in terms of its moments.

Suppose that X and Y have means μ_x and μ_y, variances σ_x^2 and σ_y^2, and covariance σ_{xy}. The problem is to derive approximations for

$E\{t(X, Y)\}$ and var $\{t(X, Y)\}$. We shall require the values of the partial derivatives

$$\frac{\partial t}{\partial x}, \frac{\partial t}{\partial y} \quad \text{and} \quad \frac{\partial^2 t}{\partial x^2}, \frac{\partial^2 t}{\partial x \partial y}, \frac{\partial^2 t}{\partial y^2},$$

at $x = \mu_x$, $y = \mu_y$. They are denoted respectively by

$$t_x, t_y \quad \text{and} \quad t_{xx}, t_{xy}, t_{yy}.$$

The approximations are based on

$$\begin{aligned} t(x, y) &\simeq t(\mu_x, \mu_y)+(x-\mu_x)t_x+(y-\mu_y)t_y \\ &\quad +\tfrac{1}{2}\{(x-\mu_x)^2 t_{xx}+2(x-\mu_x)(y-\mu_y)t_{xy}+(y-\mu_y)^2 t_{yy}\}, \\ \{t(x, y)-t(\mu_x, \mu_y)\}^2 &\simeq (x-\mu_x)^2 t_x^2+2(x-\mu_x)(y-\mu_y)t_x t_y+(y-\mu_y)^2 t_y^2. \end{aligned}$$

Averaging these expressions over the joint distribution of X and Y, we obtain

$$\begin{aligned} E\{t(X, Y)\} &\simeq t(\mu_x, \mu_y)+\tfrac{1}{2}(\sigma_x^2 t_{xx}+2\sigma_{xy}t_{xy}+\sigma_y^2 t_{yy}), \\ \text{var}\,\{t(X, Y)\} &\simeq \sigma_x^2 t_x^2+2\sigma_{xy}t_x t_y+\sigma_y^2 t_y^2. \end{aligned}$$

Example 3. The refractive index of a prism is calculated from the expression

$$t(X, Y) = \{\sin \tfrac{1}{2}(X+Y)\}/(\sin \tfrac{1}{2}X),$$

where X measures the angle of the prism and Y the minimum angular deviation of the beam of light. From the nature of the experiment, X and Y are independent variates with small variances. Consequently $E\{t(X, Y)\}$ is close to $t(\mu_x, \mu_y)$ while var $\{t(X, Y)\}$ is obtained by inserting

$$t_x = -(\sin \tfrac{1}{2}\mu_y)/(2 \sin^2 \tfrac{1}{2}\mu_x) \quad \text{and} \quad t_y = \{\cos \tfrac{1}{2}(\mu_x+\mu_y)\}/(2 \sin \tfrac{1}{2}\mu_x)$$

in the formula given.

Finally, we derive exact expressions for the mean and variance of a linear combination of variates. Suppose that $X_1, X_2, \ldots, X_n$ are variates with means $\mu_1, \mu_2, \ldots, \mu_n$ and variances $\sigma_1^2, \sigma_2^2, \ldots, \sigma_n^2$ respectively. Write

$$T = \Sigma\, a_i X_i + b,$$

where $\{a_i\}$ and b are constants. Then

$$\begin{aligned} E(T) &= \Sigma\, a_i \mu_i + b, \\ \text{var}\,(T) &= \sum_i a_i^2 \sigma_i^2 + 2 \sum_{i<j}\sum a_i a_j \,\text{cov}\,(X_i, X_j). \end{aligned}$$

These results are proved by an extension to n variates of the method used for $E\{u(X)+v(Y)\}$. Thus

$$E(\Sigma a_i X_i + b) = \Sigma a_i E(X_i) + b,$$

which proves the formula for $E(T)$. As regards the variance,

$$\begin{aligned}\text{var}(T) &= E(\Sigma a_i X_i + b - \Sigma a_i \mu_i - b)^2 \\ &= E\{\sum_i a_i^2 (X_i - \mu_i)^2 + 2 \sum_{i<j}\sum a_i a_j (X_i - \mu_i)(X_j - \mu_j)\},\end{aligned}$$

which leads at once to the form stated. When $X_1, X_2, \ldots X_n$ are mutually independent, all the covariances are zero and

$$\text{var}(T) = \Sigma a_i^2 \sigma_i^2.$$

§20. Correlation and regression

Suppose that X and Y are variates with means μ_x, μ_y and variances σ_x^2, σ_y^2 respectively. The **correlation coefficient** of X and Y is denoted by corr (X, Y) and defined by

$$\text{corr}(X, Y) = \text{cov}(X, Y)/\sigma_x \sigma_y.$$

Our results for the moments of $aX+b$ and $cY+d$ imply that corr $(aX+b, cY+d) = (ac/|ac|)$ corr $(X, Y) =$ sgn (ac) corr (X, Y), where sgn $(z) \pm 1$ according as z is positive or negative respectively. When corr (X, Y) is zero we say that X and Y are **uncorrelated**. Thus independent variates are uncorrelated.

Example 1. X and Y have a normal bivariate distribution with parameters μ_x, μ_y, σ_x^2, σ_y^2 and ρ. We have shown in the previous section that cov $(X, Y) = \rho\sigma_x\sigma_y$, and hence corr $(X, Y) = \rho$.

Example 2. Suppose that X has a density function $f(x)$ which satisfies $f(x) = f(-x)$, so that the distribution is symmetrical with $\mu = 0$. Define $Y = |X|$. Then

$$E(X) = 0, \qquad E(Y) = 2\int_0^\infty xf(x)\,dx,$$

$$E(XY) = \int_0^\infty x^2 f(x)\,dx - \int_{-\infty}^0 x^2 f(x)\,dx = 0.$$

Hence cov $(X, Y) = 0$. This example shows that two variates can be completely dependent and yet uncorrelated.

Consider the variates

$$T = (X - \mu_x)/\sigma_x \quad \text{and} \quad U = (Y - \mu_y)/\sigma_y.$$

They have zero means and unit variances, and

$$\operatorname{cov}(T, U) = E(TU) = \operatorname{corr}(T, U).$$

From the expression for the variance of a linear combination,

$$\begin{aligned}\operatorname{var}(T-U) &= \operatorname{var}(T)-2\operatorname{cov}(T, U)+\operatorname{var}(U) \\ &= 2\{1-\operatorname{corr}(T, U)\}.\end{aligned}$$

Since $\operatorname{var}(T-U) \geqslant 0$, $\operatorname{corr}(T, U) \leqslant 1$ with equality if and only if $T-U$ is constant. The constant is $E(T-U)$, which is zero. Consideration of $\operatorname{var}(T+U)$ shows in the same way that $\operatorname{corr}(T, U) \geqslant -1$ with equality if and only if $T+U = 0$. We conclude that

$$-1 \leqslant \operatorname{corr}(X, Y) \leqslant 1,$$

with equality if and only if

$$(X-\mu_x)/\sigma_x = \pm(Y-\mu_y)/\sigma_y,$$

where the $\pm$ sign is taken according as the correlation coefficient is ± 1 respectively.

These results permit some qualitative deductions about correlation. When the correlation coefficient is $+1$, the entire distribution of X and Y lies on a line of positive slope, and when the coefficient is near to $+1$, the distribution will tend to cluster about such a line. Similarly, the entire distribution lies on a line of negative slope when the correlation coefficient is -1, and lies near such a line for coefficients near to -1.

We showed in §**14** that the distribution of Y conditional on $X = x$ has density function

$$g(y \mid x) = h(x, y)/f(x).$$

The mean value of this distribution is denoted by $E(Y \mid x)$ and described as the expectation of Y conditional on $X = x$. Thus

$$E(Y \mid x) = \int yg(y \mid x)\, dy.$$

Define $E(Y \mid X)$ as the variate which takes the value $E(Y \mid x)$ with density function $f(x)$. Then

$$E\{E(Y \mid X)\} = \int f(x)\, dx \{\int yg(y \mid x)\, dy\} = \iint yh(x, y)\, dx\, dy.$$

Since the final integral is $E(Y)$, we have

$$E\{E(Y \mid X)\} = E(Y).$$

Define $XE(Y \mid X)$ as the variate which takes the value $xE(Y \mid x)$ with density function $f(x)$. A similar argument shows that

$$E\{XE(Y \mid X)\} = E(XY).$$

When x is varied, $E(Y \mid x)$ is a function of x which is known as the **regression** of Y on X. The regression is **linear** if

$$E(Y \mid x) = \theta + \beta x \quad \text{for all } x,$$

where the values of θ and β are independent of x. We say that β is a **regression coefficient**. It can be expressed in terms of the moments of X and Y as follows. Write the regression in the form

$$E(Y \mid X) = \theta + \beta X$$

and take expectations, whence

$$E(Y) = \theta + \beta E(X).$$

Thus the regression line passes through the point (μ_x, μ_y). From the last two equations

$$E(Y \mid X) - E(Y) = \beta\{X - E(X)\}.$$

Multiply each side by X and take expectations, whence

$$E(XY) - E(X)E(Y) = \beta[E(X^2) - \{E(X)\}^2].$$

Denote corr (X, Y) by ρ and we find that the slope of the regression line is

$$\beta = \rho\sigma_y/\sigma_x.$$

Hence the equation of the regression line is

$$(y - \mu_y)/\sigma_y = \rho(x - \mu_x)/\sigma_x.$$

Similarly, if the regression of X on Y is linear, then the equation of the regression line is

$$(x - \mu_x)/\sigma_x = \rho(y - \mu_y)/\sigma_y.$$

This line also passes through (μ_x, μ_y) but its slope is $\sigma_y/\rho\sigma_x$. Thus the two regression lines are distinct except when

$$\rho\sigma_y/\sigma_x = \sigma_y/\rho\sigma_x,$$

in which case $\rho^2 = 1$ and $\rho = \pm 1$. The variates X and Y then satisfy

$$(X - \mu_x)/\sigma_x = \pm(Y - \mu_y)/\sigma_y$$

and their entire distribution lies on a line, as shown earlier in the section.

Example 3. X and Y have a normal bivariate distribution with means μ_x, μ_y, variances σ_x^2, σ_y^2, and correlation coefficient ρ. According to §**15**, the distribution of Y conditional on $X = x$ is normal with mean μ and variance σ^2, where

$$\mu = \mu_y + \rho\sigma_y(x - \mu_x)/\sigma_x \quad \text{and} \quad \sigma^2 = \sigma_y^2(1-\rho^2).$$

Thus the regression of Y on X is linear and the variance of Y conditional on $X = x$ is independent of x. Similarly we find that the regression of X on Y is linear with

$$E(X \mid y) = \mu_x + \rho\sigma_x(y - \mu_y)/\sigma_y,$$

and the conditional distributions all have the same variance $\sigma_x^2(1-\rho^2)$.

EXERCISES 4

1. X has the hypergeometric distribution

$$p(X = j) = \binom{r}{j}\binom{t-r}{n-j}\bigg/\binom{t}{n},$$

where $\max(0, r+n-t) \leqslant j \leqslant \min(r, n)$. Show that

$$E(X) = nr/t \quad \text{and} \quad \text{var}(X) = nr(t-r)(t-n)/t^2(t-1).$$

Verify that the limiting values of the mean and variance when n is fixed, $t \to \infty$, and $r/t \to \theta$, are respectively $n\theta$ and $n\theta(1-\theta)$.

2. X is said to have a **lognormal** distribution if $\log X$ has a normal distribution. Suppose that $\log X$ has mean μ and variance σ^2. Show that

$$E(X) = e^{\mu + \frac{1}{2}\sigma^2} \quad \text{and} \quad \text{var}(X) = e^{2\mu}(e^{2\sigma^2} - e^{\sigma^2}).$$

3. (*a*) X has the binomial distribution

$$p(X = j) = \binom{n}{j}\theta^j(1-\theta)^{n-j} \quad (j = 0, 1, \ldots, n).$$

Show that $\text{var}\{\sin^{-1}(X/n)^{\frac{1}{2}}\} \simeq 1/4n$.

(*b*) X has the Poisson distribution

$$p(X = j) = \lambda^j e^{-\lambda}/j! \qquad (j = 0, 1, 2, \ldots).$$

Show that $\text{var}(X^{\frac{1}{2}}) \simeq 1/4$.

By considering the expansion of $n^{\frac{1}{2}} \sin^{-1}(X/n)^{\frac{1}{2}}$ when n is large, derive the result in (*b*) from that in (*a*).

4. The results of the previous exercise can be generalized as follows. Consider a variate X with mean μ and variance $\beta(\mu)$. Let $t(X)$ be such that $t''(\mu)$ is small. Given a constant c, show that $\operatorname{var}\{t(X)\} \simeq c^2$ provided that

$$t(x) = c \int^{x} \beta^{-\frac{1}{2}}(\mu)\, d\mu.$$

Hence obtain $t(X) = \sin^{-1}(X/n)^{\frac{1}{2}}$ in exercise 3(*a*) and $t(X) = X^{\frac{1}{2}}$ in exercise 3(*b*).

5. X has the density function $f(x)$ and median $\xi_{0\cdot 5}$. For any $\xi \geqslant \xi_{0\cdot 5}$, prove that

$$E\,|\,X-\xi\,| - E\,|\,X-\xi_{0\cdot 5}\,| = 2\int_{\xi_{0\cdot 5}}^{\xi} (\xi - x) f(x)\, dx.$$

State the corresponding result for $\xi \leqslant \xi_{0\cdot 5}$, and deduce that the minimum of $E\,|\,X-\xi\,|$ with respect to ξ occurs when $\xi = \xi_{0\cdot 5}$. $E\,|\,X-\xi_{0\cdot 5}\,|$ is known as the **mean deviation from the median.**

6. The joint distribution of X and Y is such that $F(0) = G(0) = 0$. Use integration by parts to show that

$$E(X) = \int_0^{\infty} \{1 - F(x)\}\, dx$$

and
$$E(XY) = \int_0^{\infty}\int_0^{\infty} \{1 - F(x) - G(y) + H(x, y)\}\, dx\, dy.$$

Deduce that

$$\operatorname{cov}(X, Y) = \int_0^{\infty}\int_0^{\infty} \{H(x, y) - F(x)G(y)\}\, dx\, dy.$$

7. Given variates Y and Z, write

$$\lambda_y = \{E(Y^2)\}^{\frac{1}{2}} \quad \text{and} \quad \lambda_z = \{E(Z^2)\}^{\frac{1}{2}}.$$

By considering $E(Y/\lambda_y \pm Z/\lambda_z)^2$, show that

$$\{E(YZ)\}^2 \leqslant E(Y^2)E(Z^2),$$

with equality if and only if $Y = cZ$ for some constant c.

8. X has mean μ and variance σ^2. Use the result of the previous exercise to show that $E\,|\,X-\mu\,| \leqslant \sigma$. Suppose also that $F(0) = 0$, in which case $|\,X^{\frac{1}{2}}\,|$ and $|\,X^{-\frac{1}{2}}\,|$ are uniquely defined. By taking $Y = |\,X^{\frac{1}{2}}\,|$ and $Z = |\,X^{-\frac{1}{2}}\,|$, prove that $E(X^{-1}) \geqslant \mu^{-1}$.

9. The distribution of Y conditional on $X = x$ has density function $g(y \mid x)$. Denote the mean and variance of this distribution by $E(Y \mid x)$ and var $(Y \mid x)$ respectively. Let var $(Y \mid X)$ be the variate which takes the value var $(Y \mid x)$ with density function $f(x)$. Prove that

$$\text{var}\,(Y) = E\{\text{var}\,(Y \mid X)\} + \text{var}\,\{E(Y \mid X)\}.$$

Verify this result in the case of a normal bivariate distribution.

10. The variates $N, X_1, X_2, \ldots$ are mutually independent and $X_1, X_2, \ldots$ all have the same distribution as X. Define

$$T = \sum_{i=1}^{N} X_i.$$

Express the mean and variance of T in terms of $E(T \mid N)$ and var $(T \mid N)$. Hence show that

$$E(T) = E(N)E(X) \quad \text{and} \quad \text{var}\,(T) = E(N)\,\text{var}\,(X) + \text{var}\,(N)\{E(X)\}^2.$$

CHAPTER 5
CALCULUS OF DISTRIBUTIONS

§21. Functions of one variate

We have described how to evaluate the mean and variance of a function of one or more variates. The exact distribution theory of such functions is essential for a precise assessment of the procedures described in later chapters. A simple version of the general problem is as follows. Let $t(x)$ be a specified function and X a variate with a known probability distribution. We require the distribution of $t(X)$.

One method consists of a direct transformation of the distribution of X. Suppose first that X is a discrete variate with the distribution

$$p(X=j)=f_j \qquad (j=0,1,2,\ldots).$$

For given k, let $j_1, j_2, \ldots$ be the solutions of the equation $t(j)=k$. Then

$$p\{t(X)=k\}=f_{j_1}+f_{j_2}+\ldots.$$

Example 1. X has a Poisson distribution with mean 1. $T=(X-1)^2$. Then $p(T=k)=0$ unless $k=j^2$ for some integer j. In particular,

$$p(T=0)=p(X=1)=\mathrm{e}^{-1},$$
$$p(T=1)=p(X=0)+p(X=2)=3\mathrm{e}^{-1}/2,$$

and for $j=2,3,\ldots,$

$$p(T=j^2)=p(X=j+1)=\mathrm{e}^{-1}/(j+1)!\,.$$

More care is required when X is a continuous variate. Let $t(x)$ be a function with a continuous positive derivative. Denote the density functions of X and $T=t(X)$ by $f(x)$ and $c(t)$ respectively. Consider an infinitesimal interval $(x, x+dx)$ and let the corresponding interval of $t(x)$ be $(t, t+dt)$. The probability elements for the two intervals are equal and therefore

$$f(x)\,dx = c(t)\,dt.$$

By the continuity of $t'(x)$, we can write $dt = t'(x)\,dx$, whence

$$c(t) = f(x)\{t'(x)\}^{-1}.$$

When $t(x)$ is a function with a continuous negative derivative, the same argument applies provided that dt is defined as the absolute value of the change in $t(x)$ corresponding to a positive increment dx, and that $t'(x)$ is replaced by $|\,t'(x)\,|$. Therefore

$$c(t) = f(x)\,|\,t'(x)\,|^{-1}$$

whenever $t(x)$ is a function with a continuous positive derivative or a continuous negative derivative.

Example 2. X is normally distributed with mean μ and variance σ^2. The corresponding density function is

$$f(x) = (2\pi\sigma^2)^{-\frac{1}{2}} \exp\{-(x-\mu)^2/2\sigma^2\}.$$

Consider the variate $T = aX+b$, where $a \neq 0$. From the results in §**17**, $E(T) = a\mu+b$ and var $(T) = a^2\sigma^2$. The function $ax+b$ has derivative a, and substitution of $x = (t-b)/a$ gives

$$\begin{aligned} c(t) &= (2\pi\sigma^2)^{-\frac{1}{2}} \exp\left\{-\left(\frac{t-b}{a}-\mu\right)^2\Big/2\sigma^2\right\}|\,a\,|^{-1} \\ &= (2\pi a^2\sigma^2)^{-\frac{1}{2}} \exp\{-(t-a\mu-b)^2/2a^2\sigma^2\}. \end{aligned}$$

Thus T is normally distributed with mean $a\mu+b$ and variance $a^2\sigma^2$, so that normal variates remain normal under linear transformations and only the mean and variance are required to specify the distribution. By taking $a = 1/\sigma$ and $b = -\mu/\sigma$, we find that $(X-\mu)/\sigma$ has a standard normal distribution.

The conditions on $t(x)$ stated above are somewhat restrictive. We relax them by assuming that $t(x)$ is a function which alternately increases and decreases i.e. the axis of x can be divided into intervals throughout which $t'(x)$ is continuous and positive, alternating with intervals throughout which $t'(x)$ is continuous and negative, together with a set of isolated points where $t'(x)$ is zero. If the equation $t(x) = t$ can be solved for x in the rth interval, denote the solution by $x_r(t)$. By following the previous argument, we find that this solution contributes $f(x_r)\,|\,t'(x_r)\,|^{-1}$ to the density function of T. Our assumptions imply that the solutions from different intervals are distinct, so that the probability elements associated with these solutions refer to mutually exclusive events. Consequently

$$c(t) = \Sigma f(x_r) \mid t'(x_r) \mid^{-1},$$

where the summation is taken over all intervals in which a solution $x_r(t)$ exists.

Now consider the expectation of T. As t varies, the solutions $x_1(t), x_2(t), \ldots$ range over separate intervals which together account for the whole of the variation in x except isolated points. Hence

$$\begin{aligned} \int tc(t)\,dt &= \int t\,\Sigma f(x_r) \mid t'(x_r) \mid^{-1} dt \\ &= \Sigma \int t(x_r) f(x_r)\,dx_r = \int t(x) f(x)\,dx. \end{aligned}$$

This shows that the alternative expressions for $E(T)$ stated in **§17** are equivalent under fairly weak conditions on $t(x)$.

Example 3. X has a standard normal distribution with density function $\phi(x)$. We require the distribution of $T = X^2$. For $t > 0$, the equation $t = x^2$ has two solutions $x = \pm\sqrt{t}$, and $\mid t'(x) \mid = 2\sqrt{t}$ at each value. Hence

$$c(t) = \{\phi(\sqrt{t}) + \phi(-\sqrt{t})\}/2\sqrt{t}.$$

Substituting for ϕ, we obtain

$$c(t) = \begin{cases} (2\pi t)^{-\frac{1}{2}} \exp(-\tfrac{1}{2}t) & (t > 0), \\ 0 & (t \leqslant 0). \end{cases}$$

The range of variation for t is $(0, \infty)$ and the corresponding intervals for $-\sqrt{t}$ and $+\sqrt{t}$ are $(-\infty, 0)$ and $(0, \infty)$ respectively, which together account for the whole of the variation in x except the point $x = 0$. We have shown in example 2 of **§17** that

$$\int_{-\infty}^{\infty} x^2 \phi(x)\,dx = 1,$$

and hence $E(T) = 1$ by this method. Alternatively,

$$E(T) = \int_0^{\infty} t(2\pi t)^{-\frac{1}{2}} \exp(-\tfrac{1}{2}t)\,dt,$$

which is transformed by the substitution $t = x^2$ into

$$2\int_0^{\infty} x^2 \phi(x)\,dx = \int_{-\infty}^{\infty} x^2 \phi(x)\,dx.$$

§22. Functions of several variates

We proceed to generalize some results of the previous section. Suppose first that X and Y are discrete variates with the joint probability distribution

$$p(X = j,\ Y = k) = h_{jk} \quad (j, k = 0, 1, 2, \ldots).$$

The distribution of $t(X, Y)$ can be obtained as follows. We enumerate the pairs of non-negative integers $(j_1, k_1), (j_2, k_2), \ldots$ which satisfy $t(j, k) = t$, and then

$$p\{t(X, Y) = t\} = h_{j_1k_1} + h_{j_2k_2} + \ldots.$$

When $t(X, Y) = X+Y$, t must be an integer, and the solutions of $j+k = t$ are $(0, t), (1, t-1), \ldots, (t, 0)$. Hence

$$p(X+Y = t) = \sum_{j=0}^{t} h_{j,t-j}.$$

If X and Y are independent, so that $h_{jk} = f_j g_k$ for all j and k, then

$$p(X+Y = t) = \sum_{j=0}^{t} f_j g_{t-j}.$$

Example 1. X and Y are independent variates with the binomial distributions

$$p(X = j) = \binom{m}{j} \theta^j (1-\theta)^{m-j} \qquad (j = 0, 1, 2, \ldots, m),$$

$$p(Y = k) = \binom{n-m}{k} \theta^k (1-\theta)^{n-m-k} \qquad (k = 0, 1, 2, \ldots, n-m).$$

The distribution of $X+Y$ is given by

$$\begin{aligned} p(X+Y = t) &= \sum_{j=0}^{t} \binom{m}{j} \theta^j (1-\theta)^{m-j} \binom{n-m}{t-j} \theta^{t-j} (1-\theta)^{(n-m)-(t-j)} \\ &= \theta^t (1-\theta)^{n-t} \sum_{j=0}^{t} \binom{m}{j}\binom{n-m}{t-j}. \end{aligned}$$

A reference to §8 shows that quantities of the form $\binom{m}{j}\binom{n-m}{t-j}\Big/\binom{n}{t}$ are the probabilities of a hypergeometric distribution, and hence

$$\sum_{j=o}^{t} \binom{m}{j}\binom{n-m}{t-j} = \binom{n}{t}.$$

Thus $X+Y$ has the binomial distribution

$$p(X+Y=t) = \binom{n}{t}\theta^t(1-\theta)^{n-t} \qquad (t = 0, 1, 2, \ldots, n).$$

Suppose next that X and Y are continuous variates with the joint density function $h(x, y)$. Let $t(x, y)$ and $u(x, y)$ be functions with continuous first-order partial derivatives, and such that the transformation from (x, y) to (t, u) is one-one. The joint density function of $T = t(X, Y)$ and $U = u(X, Y)$, say $b(t, u)$, is determined by the standard argument for changing variables in double integrals. Denote the Jacobian

$$\begin{vmatrix} \partial t/\partial x & \partial t/\partial y \\ \partial u/\partial x & \partial u/\partial y \end{vmatrix}$$

by the symbol $\dfrac{\partial(t, u)}{\partial(x, y)}$. Since the probability elements for (X, Y) and (T, U) are the same,

$$b(t, u)\,dt\,du = h(x, y)\,dx\,dy,$$

and hence

$$b(t, u) = h(x, y)\left|\frac{\partial(t, u)}{\partial(x, y)}\right|^{-1}.$$

The identity of alternative expressions for $E(T)$ follows because

$$\iint tb(t, u)\,dt\,du = \iint t(x, y)h(x, y)\,dx\,dy.$$

When the transformation specifies the functions $x(t, u)$ and $y(t, u)$ so that $t(x, y)$ and $u(x, y)$ are not explicitly defined, the Jacobian may be calculated from the result

$$\frac{\partial(t, u)}{\partial(x, y)} = \left(\frac{\partial(x, y)}{\partial(t, u)}\right)^{-1}.$$

If X and Y are independent, and $t(x)$ and $u(y)$ are each functions of one argument, then $t(X)$ and $u(Y)$ are also independent. This follows because the first condition can be expressed as

$$h(x, y) = f(x)g(y) \quad \text{for all } x \text{ and } y,$$

and the second implies that

$$\frac{\partial(t, u)}{\partial(x, y)} = \left(\frac{dt}{dx}\right)\left(\frac{du}{dy}\right).$$

Consequently

$$b(t, u) = f(x)\,|\,t'(x)\,|^{-1}g(y)\,|\,u'(y)\,|^{-1},$$

which is the product of the density functions of T and U.

We now derive the distribution of $X+Y$. Take $t = x+y$ and $u = x$. Then

$$\frac{\partial(t, u)}{\partial(x, y)} = \begin{vmatrix} 1 & 1 \\ 1 & 0 \end{vmatrix} = -1,$$

and the joint density function of $X+Y$ and X is

$$b(t, u) = h(u, t-u).$$

Thus the density function of $X+Y$ is

$$c(t) = \int h(u, t-u)\, du.$$

If X and Y are independent, we obtain

$$c(t) = \int f(x)g(t-x)\, dx.$$

The limits of integration need care if $f(x)$ or $g(y)$ vanish for some values of their arguments.

Example 2. A and B are independent standard normal variates. We shall find the distribution of $T = A^2+B^2$. According to the result given above, $X = A^2$ and $Y = B^2$ are independent. From example 3 of the previous section, the corresponding density functions are each

$$f(x) = \begin{cases} (2\pi x)^{-\frac{1}{2}} \exp(-\frac{1}{2}x) & (x > 0), \\ 0 & (x \leqslant 0). \end{cases}$$

The lower limit of the integral for $c(t)$ is zero because $f(x) = 0$ for $x \leqslant 0$, and the upper limit is t because $f(t-x) = 0$ for $x \geqslant t$. Thus

$$\begin{aligned} c(t) &= \int_0^t (2\pi x)^{-\frac{1}{2}} \exp(-\tfrac{1}{2}x)\,\{2\pi(t-x)\}^{-\frac{1}{2}} \exp\{-\tfrac{1}{2}(t-x)\}\, dx \\ &= (2\pi)^{-1} \exp(-\tfrac{1}{2}t) \int_0^t x^{-\frac{1}{2}}(t-x)^{-\frac{1}{2}}\, dx. \end{aligned}$$

Substitute $x = t \sin^2 \omega$ and the integral becomes

$$\int_0^{\frac{1}{2}\pi} (t^2 \sin^2 \omega \cos^2 \omega)^{-\frac{1}{2}}\, 2t \sin \omega \cos \omega\, d\omega = 2\int_0^{\frac{1}{2}\pi} d\omega = \pi.$$

The density function of T is therefore

$$c(t) = \tfrac{1}{2}\exp(-\tfrac{1}{2}t) \qquad (t > 0).$$

In general, let $X_1, X_2, \ldots, X_n$ have the joint density function $h(x_1, x_2, \ldots, x_n)$. We require the joint distribution of

$$T_1 = t_1(X_1, X_2, \ldots, X_n), \quad T_2 = t_2(X_1, X_2, \ldots, X_n), \quad \ldots, \quad T_r = t_r(X_1, X_2, \ldots, X_n),$$

where $r \leqslant n$. When $r < n$, further variates $T_{r+1}, \ldots, T_n$ are introduced and then eliminated in the final stage of the analysis. The assumptions are as follows.

(*i*) The set of n functions

$$t_i(x_1, x_2, \ldots, x_n) \qquad (i = 1, 2, \ldots, n)$$

have continuous first-order partial derivatives.

(*ii*) The transformation from $(x_1, x_2, \ldots, x_n)$ to $(t_1, t_2, \ldots, t_n)$ is one-one.

Denote the determinant with elements $\{\partial t_i/\partial x_j\}$ by

$$\frac{\partial(t_1, t_2, \ldots, t_n)}{\partial(x_1, x_2, \ldots, x_n)}.$$

When the transformation specifies
$x_1(t_1, t_2, \ldots, t_n), x_2(t_1, t_2, \ldots, t_n), \ldots,$
the Jacobian can be calculated from the result

$$\frac{\partial(t_1, t_2, \ldots, t_n)}{\partial(x_1, x_2, \ldots, x_n)} = \left\{\frac{\partial(x_1, x_2, \ldots, x_n)}{\partial(t_1, t_2, \ldots, t_n)}\right\}^{-1}.$$

The joint density function of $T_1, T_2, \ldots, T_n$ is

$$b(t_1, t_2, \ldots, t_n) = h(x_1, x_2, \ldots, x_n)\left|\frac{\partial(t_1, t_2, \ldots, t_n)}{\partial(x_1, x_2, \ldots, x_n)}\right|^{-1},$$

and the joint density function of $T_1, T_2, \ldots, T_r$ is obtained by integrating out $t_{r+1}, t_{r+2}, \ldots, t_n$. In the particular case where $t_1(x_1), t_2(x_2), \ldots$ are each functions of one argument, and $X_1, X_2, \ldots$ are mutually independent, we find as before that $t_1(X_1), t_2(X_2), \ldots$ are also mutually independent.

Example 3. $X_1, X_2, \ldots, X_n$ are mutually independent variates and X_j has a normal distribution with mean μ_j and variance σ_j^2

($j = 1, 2, \ldots, n$). We require the distribution of $T = \Sigma\, a_j X_j$. Define

$$Y_j = (X_j - \mu_j)/\sigma_j \qquad (j = 1, 2, \ldots, n).$$

According to example 2 of the previous section, Y_j has a standard normal distribution. The comment made above shows that $Y_1, Y_2, \ldots, Y_n$ are mutually independent. Their joint density function is therefore

$$h(y_1, y_2, \ldots, y_n) = \prod_{j=1}^{n} \{(2\pi)^{-\frac{1}{2}} \exp(-\tfrac{1}{2}y_j^2)\}$$
$$= (2\pi)^{-\frac{1}{2}n} \exp\left(-\tfrac{1}{2} \sum_{j=1}^{n} y_j^2\right).$$

Consider an orthonormal linear transformation from $y_1, y_2, \ldots, y_n$ to $t_1, t_2, \ldots, t_n$, defined by

$$\begin{aligned} t_1 &= k_{11}y_1 + k_{12}y_2 + \ldots + k_{1n}y_n, \\ t_2 &= k_{21}y_1 + k_{22}y_2 + \ldots + k_{2n}y_n, \\ &\ldots\ldots\ldots\ldots\ldots\ldots \\ t_n &= k_{n1}y_1 + k_{n2}y_2 + \ldots + k_{nn}y_n. \end{aligned}$$

The transformation can be expressed in matrix terms by

$$t = \mathscr{K} y,$$

where t is a column vector with elements $\{t_i\}$, $\mathscr{K}$ is an orthonormal matrix with elements $\{k_{ij}\}$, and y is a column vector with elements $\{y_j\}$. Since $\mathscr{K}$ is orthonormal, $\mathscr{K}'\mathscr{K} = I$, where $\mathscr{K}'$ is the transpose of $\mathscr{K}$ and I is a unit matrix. Hence

$$t't = y'\mathscr{K}'\mathscr{K} y,$$

or equivalently

$$\sum_{i=1}^{n} t_i^2 = \sum_{j=1}^{n} y_j^2.$$

Another consequence of the orthonormality of $\mathscr{K}$ is that det $\mathscr{K} = \pm 1$, whence

$$\frac{\partial(t_1, t_2, \ldots, t_n)}{\partial(y_1, y_2, \ldots, y_n)} = \pm 1.$$

Thus the joint density function of the corresponding variates $T_1, T_2, \ldots, T_n$ is

$$b(t_1, t_2, \ldots, t_n) = (2\pi)^{-\frac{1}{2}n} \exp\left(-\tfrac{1}{2} \sum_{i=1}^{n} t_i^2\right).$$

This shows that $T_1, T_2, \ldots, T_n$ are mutually independent standard normal variates. Now take

$$k_{1j} = a_j\sigma_j\{\Sigma\, a_j^2\sigma_j^2\}^{-\frac{1}{2}} \qquad (j = 1, 2, \ldots, n),$$

so that the coefficients $\{k_{1j}\}$ satisfy $\Sigma\, k_{1j}^2 = 1$. The relationships between T, $\{Y_j\}$, and T_1 imply that

$$T = \Sigma\, a_j\sigma_j Y_j + \Sigma\, a_j\mu_j = T_1\{\Sigma\, a_j^2\sigma_j^2\}^{\frac{1}{2}} + \Sigma\, a_j\mu_j.$$

Thus T has the form $aT_1 + b$. We conclude that $\Sigma\, a_j X_j$ is normally distributed with mean $\Sigma\, a_j\mu_j$ and variance $\Sigma\, a_j^2\sigma_j^2$.

§23. Distribution of χ^2

The methods of the previous section are now applied to derive one of the basic distributions of statistics. We shall need a few results concerning the gamma function, which is defined by

$$\Gamma(m) = \int_0^\infty x^{m-1}\, \mathrm{e}^{-x}\, dx \qquad (m > 0).$$

Clearly $\Gamma(1) = 1$. The value of

$$\Gamma(\tfrac{1}{2}) = \int_0^\infty x^{-\frac{1}{2}}\, \mathrm{e}^{-x}\, dx$$

is obtained as follows. Substitute $x = \frac{1}{2}t$, whence

$$\Gamma(\tfrac{1}{2}) = \int_0^\infty (2t)^{-\frac{1}{2}}\, \mathrm{e}^{-\frac{1}{2}t}\, dt.$$

According to example 3 of §**21**, $(2\pi t)^{-\frac{1}{2}}\mathrm{e}^{-\frac{1}{2}t}$ is the density function for the square of a standard normal variate, and therefore $\Gamma(\frac{1}{2}) = \pi^{\frac{1}{2}}$.

Other values of $\Gamma(m)$ can be found from the recurrence formula

$$\Gamma(m) = (m-1)\Gamma(m-1).$$

The proof consists of integrating $x^{m-1}\mathrm{e}^{-x}$ by parts. In fact,

$$\int_0^\infty x^{m-1}\mathrm{e}^{-x}\, dx = [-x^{m-1}\mathrm{e}^{-x}]_0^\infty + (m-1)\int_0^\infty x^{m-2}\mathrm{e}^{-x}\, dx,$$

and the first term on the right is zero. When m is a positive integer, say n, a repeated application of the formula gives $\Gamma(n) = (n-1)!$, because

$$\Gamma(n) = (n-1)(n-2)\ldots 1\,\Gamma(1).$$

Similarly, when m is half a positive integer, say q,

$$\Gamma(q) = (q-1)(q-2)\ldots\tfrac{1}{2}\pi^{\frac{1}{2}}.$$

We now turn to the main problem of the section. Suppose that $Y_1, Y_2, \ldots, Y_n$ are mutually independent standard normal variates, and define

$$T_1 = Y_1^2+Y_2^2+\ldots+Y_n^2.$$

We say that T_1 has a χ^2-distribution with n **degrees of freedom**, or that T_1 is distributed as χ_n^2. The density function can be obtained as follows. Introduce $U_i = Y_i^2$ for $i = 1, 2, \ldots, n$, so that

$$T_1 = U_1+U_2+\ldots+U_n.$$

The joint density function of $U_1, U_2, \ldots, U_n$ is

$$h(u_1, u_2, \ldots, u_n) = (2\pi)^{-\frac{1}{2}n}(u_1u_2\ldots u_n)^{-\frac{1}{2}}\exp\{-\tfrac{1}{2}(u_1+u_2+\ldots+u_n)\}.$$

Consider the transformation

$$u_k+u_{k+1}+\ldots+u_n = t_1t_2\ldots t_k \qquad (k = 1, 2, \ldots, n).$$

Thus $t_1 = \Sigma u_i$ and

$$t_k = (u_k+u_{k+1}+\ldots+u_n)/(u_{k-1}+u_k+\ldots+u_n) \; (k = 2, 3, \ldots, n).$$

Since $u_{k-1} > 0$, each of the variables $t_2, t_3, \ldots, t_n$ lies in (0, 1).

We first express h as a function of $t_1, t_2, \ldots, t_n$. The transformation has an inverse given by $u_n = t_1t_2\ldots t_n$ and

$$u_i = t_1t_2\ldots t_i(1-t_{i+1}) \qquad (i = 1, 2, \ldots, n-1).$$

Consequently

$$u_1u_2\ldots u_n = t_1^n\,t_2^{n-1}(1-t_2)t_3^{n-2}(1-t_3)\ldots t_n(1-t_n),$$

leading to the result

$$h(u_1, u_2, \ldots, u_n) = (2\pi)^{-\frac{1}{2}n}\{t_1^n\,t_2^{n-1}(1-t_2)t_3^{n-2}(1-t_3)\ldots t_n(1-t_n)\}^{-\frac{1}{2}}e^{-\frac{1}{2}t_1}.$$

We evaluate the Jacobian by dividing the transformation into two stages, first from $\{u_i\}$ to $\{w_j\}$, and then from $\{w_j\}$ to $\{t_k\}$, where

$$w_j = u_j+u_{j+1}+\ldots+u_n \qquad (j = 1, 2, \ldots, n).$$

$\partial(w_1, w_2, \ldots, w_n)/\partial(u_1, u_2, \ldots, u_n)$ has zero elements below the main diagonal and its value is the product of the diagonal elements, which are all 1. Hence also

$$\frac{\partial(u_1, u_2, \ldots, u_n)}{\partial(w_1, w_2, \ldots, w_n)} = 1.$$

In the second stage

$$w_k = t_1 t_2 \ldots t_k \qquad (k = 1, 2, \ldots, n).$$

$\partial(w_1, w_2, \ldots, w_n)/\partial(t_1, t_2, \ldots, t_n)$ has zero elements above the main diagonal, and its value is the product of the diagonal elements, so that

$$\frac{\partial(w_1, w_2, \ldots, w_n)}{\partial(t_1, t_2, \ldots, t_n)} = 1(t_1)(t_1 t_2) \ldots (t_1 t_2 \ldots t_{n-1}).$$

The value of $\partial(u_1, u_2, \ldots, u_n)/\partial(t_1, t_2, \ldots, t_n)$ is therefore

$$\frac{\partial(u_1, u_2, \ldots, u_n)}{\partial(w_1, w_2, \ldots, w_n)} \frac{\partial(w_1, w_2, \ldots, w_n)}{\partial(t_1, t_2, \ldots, t_n)} = t_1^{n-1}\, t_2^{n-2} \ldots t_{n-1}.$$

Combining the end results of the two previous paragraphs, we find that the joint density function of $T_1, T_2, \ldots, T_n$ is

$$b(t_1, t_2, \ldots, t_n)$$
$$= (2\pi)^{-\frac{1}{2}n} t_1^{\frac{1}{2}(n-2)} t_2^{\frac{1}{2}(n-3)} (1-t_2)^{-\frac{1}{2}} \ldots t_n^{-\frac{1}{2}} (1-t_n)^{-\frac{1}{2}} e^{-\frac{1}{2}t_1}.$$

When $t_2, t_3, \ldots, t_n$ are integrated out, what remains is proportional to $t_1^{\frac{1}{2}n-1} e^{-\frac{1}{2}t_1}$. The density function of T_1, say q_n, is therefore given by

$$q_n(t_1) = t_1^{\frac{1}{2}n-1} e^{-\frac{1}{2}t_1} \Big/ \int_0^\infty t_1^{\frac{1}{2}n-1} e^{-\frac{1}{2}t_1} dt_1.$$

Substitute $t_1 = 2x$ in the integral, which then becomes

$$2^{\frac{1}{2}n} \int_0^\infty x^{\frac{1}{2}n-1} e^{-x}\, dx = 2^{\frac{1}{2}n}\Gamma(\tfrac{1}{2}n).$$

Thus the density function of χ_n^2 is

$$q_n(t) = t^{\frac{1}{2}n-1} e^{-\frac{1}{2}t}/2^{\frac{1}{2}n}\Gamma(\tfrac{1}{2}n). \qquad (t > 0).$$

Some representative χ^2-distributions are shown in Fig. 10. When $n = 1$, $q_1(t) \to \infty$ as $t \to 0$. When $n = 2$, the distribution is exponential with parameter $\frac{1}{2}$, as already shown in example 2 of the previous

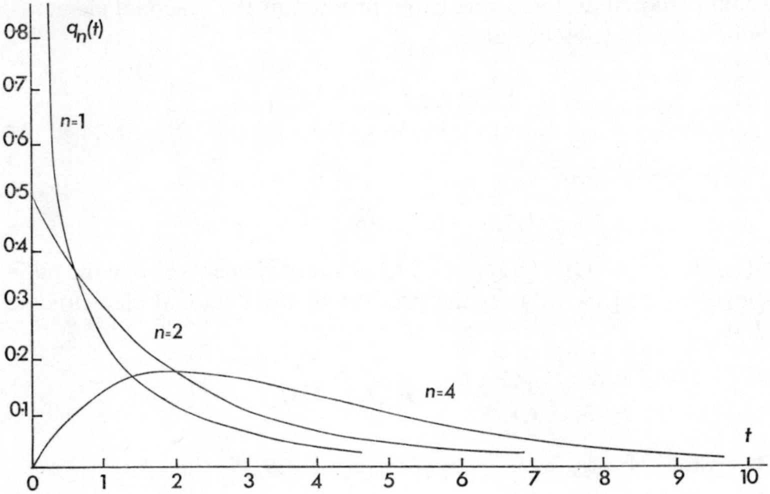

Fig. 10. Density function of χ^2 with 1, 2, and 4 degree of freedom.

section. For $n > 2$, the density function rises to a maximum as t increases and then falls off more gradually.

Suppose that X_1 and X_2 are independent variates which are distributed as $\chi^2_{n_1}$ and $\chi^2_{n_2}$ respectively. We shall prove that X_1+X_2 is distributed as $\chi^2_{n_1+n_2}$. The joint density function of X_1 and X_2 is

$$x_1^{\frac{1}{2}n_1-1}x_2^{\frac{1}{2}n_2-1}e^{-\frac{1}{2}(x_1+x_2)} \,/\, 2^{\frac{1}{2}(n_1+n_2)}\Gamma(\tfrac{1}{2}n_1)\Gamma(\tfrac{1}{2}n_2).$$

Make a transformation of the type already used in deriving the χ^2-distribution, namely

$$x_1+x_2 = v_1 \quad \text{and} \quad x_2 = v_1v_2.$$

Then $x_1 = v_1(1-v_2)$ and

$$\frac{\partial(x_1, x_2)}{\partial(v_1, v_2)} = v_1.$$

The joint density function of V_1 and V_2 is therefore

$$v_1^{\frac{1}{2}(n_1+n_2)-1}v_2^{\frac{1}{2}n_2-1}(1-v_2)^{\frac{1}{2}n_1-1}e^{-\frac{1}{2}v_1} \,/\, 2^{\frac{1}{2}(n_1+n_2)}\Gamma(\tfrac{1}{2}n_1)\Gamma(\tfrac{1}{2}n_2).$$

This shows that the density function of V_1 is proportional to $v_1^{\frac{1}{2}(n_1+n_2)-1}e^{-\frac{1}{2}v_1}$, which has the form characteristic of χ^2. Thus $X + X_2$ is distributed as $\chi^2_{n_1+n_2}$. By induction, if $X_1, X_2, \ldots$ are mutually independent variates which are distributed as $\chi^2_{n_1}, \chi^2_{n_2}, \ldots$ respectively, then ΣX_i is distributed as $\chi^2_{\Sigma n_i}$.

§24. Probability generating functions

Let X be a variate with the probability distribution

$$p(X = j) = f_j \qquad (j = 0, 1, 2, \ldots).$$

The **probability generating function** (p.g.f.) of X is denoted by $P(z)$ and defined by

$$P(z) = E(z^X) = \sum_{j=0}^{\infty} f_j z^j.$$

Since $P(1) = 1$, the series converges for $|z| \leqslant 1$, but it may also converge for larger values of z. Suppose that $P(1+y)$ converges in a neighbourhood of $y = 0$. The coefficient of y^r in $P(1+y)$ is

$$\sum_{j=r}^{\infty} f_j \binom{j}{r} = \sum_{j=r}^{\infty} j(j-1)\ldots(j-r+1)f_j/r! = \mu_{(r)}/r!,$$

and consequently

$$P(1+y) = \sum_{r=0}^{\infty} \mu_{(r)} y^r/r!.$$

We therefore say that $P(1+y)$ is the **factorial moment generating function** of X. The factorial moments can also be found by differentiating $P(1+y)$ at $y = 0$ or, what is the same thing, differentiating $P(z)$ at $z = 1$. In this way, $\mu_{(r)} = P^{(r)}(1)$, and relations with the central moments imply that

$$E(X) = P'(1) \quad \text{and} \quad \operatorname{var}(X) = P''(1) + P'(1) - \{P'(1)\}^2.$$

Example 1. X has the binomial distribution

$$p(X = j) = \binom{n}{j} \theta^j (1-\theta)^{n-j} \qquad (j = 0, 1, \ldots, n).$$

The probability generating function is

$$P(z) = \sum_{j=0}^{n} \binom{n}{j} \theta^j (1-\theta)^{n-j} z^j = \{\theta z + (1-\theta)\}^n.$$

Hence $P(1+y) = (1+\theta y)^n$ and the factorial moments are

$$\mu_{(r)} = n(n-1)\ldots(n-r+1)\theta^r \qquad (r = 1, 2, \ldots, n).$$

Write $\theta = \lambda/n$ and take the limit of $P(z)$ as $n \to \infty$. We obtain

$$\lim_{n \to \infty} \{1+\lambda(z-1)/n\}^n = \exp\{\lambda(z-1)\}.$$

The coefficient of z^j in $\exp\{\lambda(z-1)\}$ is $\lambda^j e^{-\lambda}/j!$. Thus $\exp\{\lambda(z-1)\}$ is the p.g.f. of a Poisson distribution with parameter λ. We have seen in §**6** that the limit of a binomial distribution under the conditions stated is a Poisson distribution.

The concept of a probability generating function is extended to two variates as follows. Let X and Y have the probability generating functions

$$P_x(z) = E(z^X) \quad \text{and} \quad P_y(w) = E(w^Y).$$

Suppose that the joint probability distribution of X and Y is given by

$$p(X = j, Y = k) = h_{jk} \qquad (j, k = 0, 1, 2, \ldots).$$

Their joint probability generating function $Q(z, w)$ is defined by

$$Q(z, w) = E(z^X w^Y) = \sum_{j=0}^{\infty} \sum_{k=0}^{\infty} h_{jk} z^j w^k.$$

Since $Q(1, 1) = 1$, the double series converges absolutely for $|z| \leqslant 1$ and $|w| \leqslant 1$. By taking $w = 1$ or $z = 1$, we find that

$$P_x(z) = Q(z, 1) \quad \text{and} \quad P_y(w) = Q(1, w).$$

The probability generating function of $X+Y$ is

$$E(z^{X+Y}) = Q(z, z).$$

If X and Y are independent, then the functions z^X and w^Y are also independent, so that

$$E(z^X w^Y) = E(z^X)E(w^Y).$$

We therefore obtain

$$Q(z, w) = P_x(z)\,P_y(w).$$

Conversely, this condition on the probability generating functions can be written in the form

$$\Sigma\,\Sigma\,(h_{jk}-f_j g_k) z^j w^k = 0,$$

where $\{f_j\}$ and $\{g_k\}$ are the distributions of X and Y respectively.

The double series is absolutely convergent for all z and w such that $|z| \leqslant 1$ and $|w| \leqslant 1$. Hence each coefficient is zero, and

$$h_{jk} = f_j g_k \quad \text{for all } j \text{ and } k.$$

Thus a necessary and sufficient condition for the independence of X and Y is that $Q(z, w) = P_x(z)\,P_y(w)$ for all z and w such that $|z| \leqslant 1$ and $|w| \leqslant 1$.

Example 2. X and Y are independent Poisson variates with parameters λ and μ respectively. The p.g.f. of $X+Y$ is

$$P_x(z)\,P_y(z) = e^{\lambda(z-1)}e^{\mu(z-1)} = e^{(\lambda+\mu)(z-1)},$$

which shows that $X+Y$ has a Poisson distribution with parameter $\lambda+\mu$.

Further generalizations are straightforward. The joint probability generating function of $X_1, X_2, \ldots, X_n$ is given by

$$R(z_1, z_2, \ldots, z_n) = E(z_1^{X_1} z_2^{X_2} \ldots z_n^{X_n}).$$

A useful result is the following. Let $X_1, X_2, \ldots, X_n$ be mutually independent variates with probability generating functions $P_1(z)$, $P_2(z), \ldots, P_n(z)$ respectively. Then the p.g.f. of ΣX_i is $\Pi P_i(z)$. In fact, $z^{X_1}, z^{X_2}, \ldots, z^{X_n}$ are mutually independent, whence

$$E(z^{\Sigma X_i}) = \Pi E(z^{X_i}).$$

§25. Moment generating functions

Let X be a continuous variate with density function $f(x)$, and define

$$J(z) = E(e^{zX}) = \int_{-\infty}^{\infty} e^{zx} f(x)\,dx.$$

Clearly $J(0) = 1$. Suppose that $J(z)$ exists for all values of z in a real interval containing $z = 0$. We can then expand $J(z)$ in the form

$$J(z) = \sum_{r=0}^{\infty} J^{(r)}(0)\,z^r/r!,$$

where $J^{(r)}(0)$ stands for the rth derivative of $J(z)$ evaluated at $z = 0$. From the definition of $J(z)$,

$$J^{(r)}(0) = \int_{-\infty}^{\infty} x^r f(x)\,dx = \mu_r'.$$

Consequently

$$J(z) = \sum_{r=0}^{\infty} \mu'_r z^r / r!.$$

Thus μ'_r is the coefficient of $z^r/r!$ when $J(z)$ is expanded in a power series. We therefore say that $J(z)$ is the **moment generating function** (m.g.f.) of X.

The moment generating function of $T = aX+b$ is

$$E\{e^{z(aX+b)}\} = e^{bz}E\{e^{(az)X}\}.$$

We can write the result as

$$J_t(z) = e^{bz}J_x(az),$$

where J_t and J_x are the moment generating functions of T and X respectively. In particular, the m.g.f. of $(X-\mu)$ is $e^{-\mu z}J_x(z)$. On the other hand

$$E\{e^{z(X-\mu)}\} = E\left\{\sum_{r=0}^{\infty} z^r(X-\mu)^r/r!\right\} = \sum_{r=0}^{\infty} \mu_r z^r/r!\,.$$

This shows that the central moments of X can be obtained from $e^{-\mu z}J_x(z)$ by expansion in series or repeated differentiation at $z = 0$.

The assumption that $J(z)$ exists throughout a real interval containing the origin holds for many of the distributions which arise in statistical practice, but there are other distributions for which the assumption fails to hold. This difficulty is met by considering the function $D(q) = E(e^{iqX})$, where $i = \sqrt{(-1)}$ and q is real. $D(q)$ always exists because

$$|\,D(q)\,| \leqslant \int_{-\infty}^{\infty} |\,e^{iqx}\,|\,f(x)\,dx = 1.$$

We say that $D(q)$ is the **characteristic function** of X, because there is a unique distribution corresponding to a given characteristic function. A general proof that the characteristic function uniquely determines the probability distribution belongs to the theory of integral transforms and lies beyond the scope of this book, but we shall find it convenient to use the result occasionally. The fact that a distribution is identified by its characteristic function provides an alternative approach to problems which are difficult or impossible to solve by the methods of §**22**.

Example 1. X has a normal distribution with mean μ and variance σ^2. Consider the variate $T = (X-\mu)/\sigma$. According to example 1 of §**21**, T has a standard normal distribution. Its moment generating function is

$$J_t(z) = \int_{-\infty}^{\infty} (2\pi)^{-\frac{1}{2}} \exp(zt - \tfrac{1}{2}t^2)\, dt.$$

The term inside the exponential is

$$zt - \tfrac{1}{2}t^2 = \tfrac{1}{2}z^2 - \tfrac{1}{2}(t-z)^2.$$

Consequently

$$J_t(z) = \exp(\tfrac{1}{2}z^2) \int_{-\infty}^{\infty} (2\pi)^{-\frac{1}{2}} \exp\{-\tfrac{1}{2}(t-z)^2\}\, dt.$$

The integrand is the density function of a normal distribution with mean z and variance 1. Hence the integral is unity and

$$J_t(z) = \exp(\tfrac{1}{2}z^2).$$

The moment generating function of $X = \sigma T + \mu$ is thus

$$J_x(z) = e^{\mu z} J_t(\sigma z) = \exp(\mu z + \tfrac{1}{2}\sigma^2 z^2).$$

We obtain the central moments of X by expanding

$$e^{-\mu z} J_x(z) = \exp(\tfrac{1}{2}\sigma^2 z^2).$$

For instance, μ_4 is the coefficient of $z^4/4!$, namely

$$4!\,(\tfrac{1}{2}\sigma^2)^2/2! = 3\sigma^4.$$

Now let X and Y be continuous variates with the moment generating functions

$$J_x(z) = E(e^{zX}) \quad \text{and} \quad J_y(w) = E(e^{wY}).$$

Denote the joint density function of X and Y by $h(x, y)$ and define

$$K(z, w) = E(e^{zX+wY}) = \int_{-\infty}^{\infty} \int_{-\infty}^{\infty} e^{zx+wy} h(x, y)\, dx\, dy.$$

Suppose that K exists throughout a region of the (z, w) plane containing the origin, in which case

$$K(z, w) = \sum_{r=0}^{\infty} \sum_{s=0}^{\infty} z^r w^s E(X^r Y^s)/r!\, s!\,.$$

We say that $K(z, w)$ is the joint moment generating function of X and Y. The moment generating functions of X and Y individually are obtained by taking $w = 0$ and $z = 0$ respectively. Thus

$$J_x(z) = K(z, 0) \quad \text{and} \quad J_y(w) = K(0, w).$$

The joint m.g.f. of $T = aX+b$ and $U = cY+d$ is

$$E\{e^{z(aX+b)+w(cY+d)}\} = e^{bz+dw}E\{e^{azX+cwY}\}.$$

This can be expressed in the form

$$K_{t,u}(z, w) = e^{bz+dw}K_{x,y}(az, cw).$$

If X and Y are independent, then so are the functions e^{zX} and e^{wY}, whence

$$K(z, w) = J_x(z)J_y(w).$$

Suppose that $z = iq$ and $w = ip$ where $i = \sqrt{(-1)}$ and q, p are real. The condition that

$$E(e^{iqX+ipY}) = D_x(q)\,D_y(p)$$

for all q and p implies that X and Y are independent, but the theory of integral transforms is necessary in order to prove this result.

Example 2. X and Y have a normal bivariate distribution with means μ_x, μ_y, variances σ_x^2, σ_y^2 and correlation coefficient ρ. We require $K(z, w)$. Consider first the joint distribution of $T = (X-\mu_x)/\sigma_x$ and $U = (Y-\mu_y)/\sigma_y$. The Jacobian is

$$\partial(t, u)/\partial(x, y) = 1/\sigma_x\sigma_y.$$

Hence the joint density function of T and U is

$$h(x, y)\sigma_x\sigma_y = (2\pi)^{-1}(1-\rho^2)^{-\frac{1}{2}}\exp\{-(t^2-2\rho tu+u^2)/2(1-\rho^2)\}.$$

Thus T and U have a standard normal bivariate distribution with joint density function $\psi(t, u)$. Their joint moment generating function is

$$K_{t,u}(z, w) = \int_{-\infty}^{\infty}\int_{-\infty}^{\infty} e^{zt+wu}\psi(t, u)\,dt\,du.$$

As before, we use the factorization

$$\psi(t, u) = \phi(t)\{2\pi(1-\rho^2)\}^{-\frac{1}{2}}\exp\{-(u-\rho t)^2/2(1-\rho^2)\},$$

where ϕ is the standard normal density function. This gives

$$K_{t,u}(z, w) = \int_{-\infty}^{\infty} e^{zt}\phi(t)\,dt\,[\int_{-\infty}^{\infty} e^{wu}\{2\pi(1-\rho^2)^{-\frac{1}{2}}\}$$
$$\exp\{-(u-\rho t)^2/2(1-\rho^2)\}\,du].$$

The integral in square brackets is the m.g.f. of a normal distribution with mean ρt and variance $(1-\rho^2)$. From example 1, the value of this integral is $\exp\{\rho tw+\frac{1}{2}(1-\rho^2)w^2\}$. Consequently

$$K_{t,u}(z, w) = \exp\{\tfrac{1}{2}(1-\rho^2)w^2\} \int_{-\infty}^{\infty} e^{(z+\rho w)t}\phi(t)\,dt.$$

The integral here is $\exp\{\frac{1}{2}(z+\rho w)^2\}$, and hence

$$K_{t,u}(z, w)=\exp\{\tfrac{1}{2}(1-\rho^2)w^2+\tfrac{1}{2}(z+\rho w)^2\}=\exp\{\tfrac{1}{2}(z^2+2\rho zw+w^2)\}.$$

Finally, the joint moment generating function of $X = \sigma_x T+\mu_x$ and $Y = \sigma_y U+\mu_y$ is

$$\begin{aligned}K_{x,y}(z, w) &= e^{\mu_x z+\mu_y w}K_{t,u}(\sigma_x z, \sigma_y w)\\ &= \exp\{\mu_x z+\mu_y w+\tfrac{1}{2}(\sigma_x^2 z^2+2\rho\sigma_x\sigma_y zw+\sigma_y^2 w^2)\}.\end{aligned}$$

We briefly note the main result for n variates. Let $X_1, X_2, \ldots, X_n$ be mutually independent and denote their moment generating functions by $J_1(z), J_2(z), \ldots, J_n(z)$ respectively. Then the m.g.f. of ΣX_i is $\Pi J_i(z)$. The proof is the same as for probability generating functions.

Example 3. $X_1, X_2, \ldots, X_n$ are mutually independent variates and X_i has a normal distribution with mean μ_i and variance σ_i^2 $(i = 1, 2, \ldots, n)$. We require the distribution of $\Sigma a_i X_i$. Since $a_i X_i$ is normal with mean $a_i\mu_i$ and variance $a_i^2\sigma_i^2$, its moment generating function is

$$J_i(z) = \exp(a_i\mu_i z+\tfrac{1}{2}a_i^2\sigma_i^2 z^2).$$

Hence the moment generating function of $\Sigma a_i X_i$ is

$$\Pi J_i(z) = \exp(z\Sigma a_i\mu_i+\tfrac{1}{2}z^2\Sigma a_i^2\sigma_i^2).$$

This is the m.g.f. of a normal variate with mean $\Sigma a_i\mu_i$ and variance $\Sigma a_i^2\sigma_i^2$. We can prove by integration in the complex plane that the corresponding characteristic function is obtained on replacing z by $q\sqrt{(-1)}$. An appeal to the uniqueness theorem then leads to the conclusion that $\Sigma a_i X_i$ is normal with mean $\Sigma a_i\mu_i$ and variance $\Sigma a_i^2\sigma_i^2$. The same problem was solved at greater length in §**21**, where all the details were given.

EXERCISES 5

1. X is uniformly distributed on $(0, \pi)$. Prove that the density function of $T = a \cos X + b$ is

$$c(t) = 1/\pi\{a^2 - (t-b)^2\} \qquad (b-a < t < b+a).$$

2. T is said to have a **Cauchy** distribution with parameters α and β if its density function is $\beta/\pi\{\beta^2 + (t-\alpha)^2\}$, where $-\infty < \alpha < \infty$, and $\beta > 0$. Show that T has median α but that $E(T)$ does not exist. Find the distribution function of T.

If X and Y have a standard normal bivariate distribution, prove that X/Y has a Cauchy distribution with parameters ρ and $(1-\rho^2)^{\frac{1}{2}}$.

3. $X_1, X_2, \ldots, X_n$ are mutually independent, and each has an exponential distribution with parameter λ. Prove by induction that the density function of $T = \Sigma X_i$ is

$$c(t) = \lambda^n t^{n-1} \mathrm{e}^{-\lambda t}/(n-1)! \qquad (t > 0).$$

Verify the result by the use of moment generating functions.

4. $X_1, X_2, \ldots, X_n$ are mutually independent and each is uniformly distributed on $(0, 1)$. Show that $-\log X_i$ has an exponential distribution with parameter 1. Hence prove that $-2 \log \Pi X_i$ is distributed as χ^2_{2n}.

5. Consider molecules moving at random in a container and let X_1, X_2, X_3 be the components of the velocity of a molecule along three mutually orthogonal axes. Suppose that X_1, X_2, X_3 are mutually independent standard normal variates. Show that the density function of the magnitude of velocity is

$$c(t) = (2/\pi)^{\frac{1}{2}} t^2 \mathrm{e}^{-\frac{1}{2}t^2} \qquad (t > 0).$$

This is known as **Maxwell's** distribution.

6. $X_1, X_2, \ldots, X_k$ have a multinomial distribution

$$p(X_1 = x_1, \ldots, X_k = x_k) = n!\,\theta_1^{x_1} \ldots \theta_k^{x_k}/x_1! \ldots x_k!,$$

where $x_i \geqslant 0$, $\Sigma x_i = n$, and $\Sigma\theta_i = 1$. Prove that the joint probability generating function of $X_1, X_2, \ldots, X_k$ is

$$R(z_1, \ldots, z_k) = (\theta_1 z_1 + \ldots + \theta_k z_k)^n.$$

Hence show that $\operatorname{cov}(X_i, X_j) = -n\theta_i\theta_j \quad (i \neq j)$.

7. $X_1, X_2, \ldots, X_n$ are mutually independent and each has the negative binomial distribution.

$$p(X = j) = \binom{b+j-1}{b-1}(1-\theta)^b\theta^j \qquad (j = 0, 1, 2, \ldots),$$

where b is a positive integer and $0 < \theta < 1$. Prove that the distribution of $T = \Sigma X_i$ has the same form with b replaced by nb. Show that

$$E(T) = nb\theta/(1-\theta) \quad \text{and} \quad \text{var}(T) = nb\theta/(1-\theta)^2.$$

8. X and Y have a normal bivariate distribution. Show that $aX+bY$ is normally distributed.

9. The purchase of one or more Premium Savings Bonds entitles the holder to participate in future monthly draws. A single bond wins a prize in any draw with probability θ. The prize money is a variate X, where X is a positive integer. Given that a prize is won, the probability generating function of prize money is $P(z)$. A person holds b bonds and participates in m draws. Show that the probability generating function of his total prize money is

$$\{1-\theta+\theta P(z)\}^{bm}.$$

10. $X_1, X_2, \ldots, X_n$ are mutually independent and X_i is normally distributed with mean μ_i and variance 1 $(i = 1, 2, \ldots, n)$. Prove that the moment generating function of X_i^2 is

$$J(z) = (1-2z)^{-\frac{1}{2}} \exp\{-\tfrac{1}{2}\mu_i^2 + \tfrac{1}{2}\mu_i^2/(1-2z)\}.$$

Hence show that the density function of ΣX_i^2 is

$$c(t) = \exp(-\tfrac{1}{2}\eta) \sum_{k=0}^{\infty} (\tfrac{1}{2}\eta)^k q_{n+2k}(t)/k!,$$

where $\eta = \Sigma\mu_i^2$ and q_m is the density function of χ_m^2. We say that ΣX_i^2 has a **non-central** χ^2-distribution with n degrees of freedom and parameter η.

CHAPTER 6
RANDOM SAMPLES

§26. Random samples

Several of the probability distributions discussed in chapters 2 and 3 were illustrated by sets of data collected under fixed experimental conditions. We deferred technical comment on comparisons of the frequencies provided by observation and theory, but methods of the type to be developed in chapter 11 will confirm that there is satisfactory agreement in every case. This result suggests that each set of data can be regarded as consisting of repeated observations on the values taken by a particular variate e.g. that the intervals between emission of α-particles were obtained as measurements of a variate with an exponential distribution. When observations on a variate X can be repeated indefinitely, the frequency-ratio for any specified range of values will tend to the corresponding probability. The set of observations approached in the limit is known as a **population**, and it serves to represent the probability distribution of X in a form that can be readily imagined. A rigorous definition is unnecessary because the population is essentially equivalent to the distribution of X. Thus we refer to $E(X)$ as the **population mean**, and similarly for other characteristics of the distribution.

Only a finite set of observations on X is available in practice, and this set is described as a sample taken from the distribution of X. The object of sampling is to acquire or increase knowledge of the distribution. When the distribution is completely specified except for the values of one or more parameters, the object is best achieved if each parameter is estimated by a suitable function of the sample observations. The sample mean has already been introduced and other quantities calculated from the sample will appear in the course of this chapter. Any function of the sample observations is described as a **statistic**. The properties of a statistic can be determined only when the mathematical structure of the observations is clearly defined, and the concept of a **random sample** provides a simple framework which is appropriate whenever the data are obtained from independent identical experiments.

Suppose that $X_1, X_2, \ldots, X_n$ are mutually independent variates, each with the same distribution as X. A random sample of size n taken from the probability distribution of X consists of n values $x_1, x_2, \ldots, x_n$ which are observations on $X_1, X_2, \ldots, X_n$ respectively. Any function t of n variables is a statistic for samples of size n, and each sample provides one value $t(x_1, x_2, \ldots, x_n)$ which is an observation on the variate $t(X_1, X_2, \ldots, X_n)$. The probability distribution of $t(X_1, X_2, \ldots, X_n)$ is known as the sampling distribution of the statistic t. Thus we refer to the distribution of $\Sigma X_i/n$ as the distribution of the sample mean.

The concept of a random sample is readily extended to bivariate and multivariate situations. For example, suppose that the continuous variates X and Y have the joint density function $h(x, y)$, and that $(X_1, Y_1), (X_2, Y_2), \ldots, (X_n, Y_n)$ each have the same distribution as (X, Y). Then $(X_1, Y_1), (X_2, Y_2), \ldots, (X_n, Y_n)$ are described as mutually independent pairs of variates if the joint density function of the $2n$ variates $X_1, Y_1, X_2, Y_2, \ldots, X_n, Y_n$ is given by

$$h(x_1, y_1)\, h(x_2, y_2) \ldots h(x_n, y_n).$$

A random sample of size n from the joint distribution of X and Y consists of n pairs $(x_1, y_1), (x_2, y_2), \ldots, (x_n, y_n)$, which are observations on $(X_1, Y_1), (X_2, Y_2), \ldots, (X_n, Y_n)$ respectively.

§27. Random sampling numbers

Random samples can be constructed artificially and then used to make an empirical study of the properties of any statistic. The simplest method of construction is based on a table of **random sampling numbers**. This is a list of decimal digits in haphazard order, and can be regarded as a sequence of independent observations on a variate D with a discrete uniform distribution, defined by

$$p(D = d) = 1/10 \qquad (d = 0, 1, 2, \ldots, 9).$$

The validity of a table of random sampling numbers is established by the application of tests which are based on the properties of a sequence of independent variates each distributed like D. A valid table is given by Fisher and Yates (1963), and others are listed by Greenwood and Hartley (1962). When a computer is available there are standard programs which generate **pseudo-random numbers**. A finite sequence of pseudo-random numbers, from a program which has been carefully tested, can be regarded as a set of independent observations on a variate U with a continuous uniform distribution, defined by the density function

$$f(u) = 1 \qquad (0 < u < 1).$$

The reason for the prefix pseudo- is that the numbers are produced in a deterministic way and will eventually recur in the same order, so that the sequence is not properly random. However, the program can be chosen so that the period of recurrence is very long, and thus there is no restriction on the use of pseudo-random numbers in practice.

A sequence of b observations on D forms the first b digits in the decimal expansion of an observation on U, and thus we can use a table of random sampling numbers to generate what is approximately a sample of uniform variates. On the other hand, a consequence of the usual method of generation of pseudo-random numbers is that the digits obtained from them are not satisfactorily random.

A set of random or pseudo-random numbers is used in the following way to construct a random sample. Suppose that X is a discrete variate with the distribution

$$p(X = j) = f_j \qquad (j = 0, 1, 2, \ldots).$$

We calculate the cumulative probabilities $f_0, f_0+f_1, f_0+f_1+f_2, \ldots$ and assign a value j to the sample whenever the observation on U is such that

$$f_0+f_1+\ldots+f_{j-1} < u < f_0+f_1+\ldots+f_j.$$

The probability that U lies in this interval is f_j and hence the correct probability is assigned to $X = j$. When X is a continuous variate with distribution function $F(x)$, we divide the axis of x into equal intervals of length h, say at the points

$$\ldots, k-1\tfrac{1}{2}h,\ k-\tfrac{1}{2}h,\ k+\tfrac{1}{2}h,\ k+1\tfrac{1}{2}h, \ldots,$$

and approximate X by a discrete variate Y with the distribution

$$p(Y = k) = F(k+\tfrac{1}{2}h)-F(k-\tfrac{1}{2}h).$$

The previous procedure is then applied, so that k is a sample member whenever

$$F(k-\tfrac{1}{2}h) < u < F(k+\tfrac{1}{2}h).$$

Example. X has a standard normal distribution, with distribution function $\Phi(x)$. A random sample is required where the sample values are recorded to one decimal place. We give below an extract from a table of $\Phi(x)$. The values of x are such that the centre of each x-interval is a multiple of 0·1.

x	$\Phi(x)$	x	$\Phi(x)$
...	...	...	...
−1·45	0·0735	0·05	0·5199
−1·35	0·0885	0·15	0·5596
−1·25	0·1056	0·25	0·5987
...	...	...	...

We now take from a table of random sampling numbers just enough digits to identify the value of k. Thus 54 gives $k = 0{\cdot}1$ and 0887 gives $k = -1{\cdot}3$.

We say that r random samples of size n from the distribution of X are independent if the rn variates observed are mutually independent. Each sample provides one observation on the variate $T = t(X_1, X_2, \ldots, X_n)$. A frequency distribution of r observations on T can be formed by taking r independent random samples of size n, and a satisfactory approximation to the probability distribution of T is usually obtained when r is about 500 or 1000. The whole procedure consists of generating observations on U, converting them into observations on X, calculating $t(x_1, x_2, \ldots, x_n)$ from each block of n observations, and finally sorting the values of t into intervals of suitable length. All these operations can be rapidly carried out on a computer, but they are otherwise distinctly laborious. This method of approximating to a probability distribution is known as **simulation**, and it can provide knowledge about the distribution of a statistic in circumstances where the techniques of the previous chapter fail to give explicit information.

§28. Sample moments

A random sample of size n is taken from the distribution of X. The sample values $x_1, x_2, \ldots, x_n$ are observations on $X_1, X_2, \ldots, X_n$ respectively. Consider first the statistic

$$m = \Sigma x_i/n.$$

This is an observation on the variate

$$M = \Sigma X_i/n.$$

Both m and M are described as the sample mean, and the context will make clear whether the observation or variate is being discussed. Suppose that X has mean μ and variance σ^2, both finite. Then $X_1, X_2, \ldots, X_n$ each have the same mean and variance as X, and the members of any pair (X_i, X_j) are uncorrelated. Using the results in §**19** concerning the mean and variance of a linear combination of variates, we find that

$$E(M) = \Sigma E(X_i)/n = \mu,$$

and

$$\text{var}\,(M) = \Sigma\,\text{var}\,(X_i)/n^2 = \sigma^2/n.$$

When a statistic t is used to estimate a parameter θ, we say that $t(X_1, X_2, \ldots, X_n)$ is an **estimator** of θ and $t(x_1, x_2, \ldots, x_n)$ is an **estimate** of θ. The estimator is **unbiased** if

$$E\{t(X_1, X_2, \ldots, X_n)\} = \theta,$$

and otherwise, $t(X_1, X_2, \ldots, X_n)$ is a **biased** estimator of θ. The sample mean and any sample member are unbiased estimators of μ, but the sample mean is preferable because its variance is smaller by a factor n^{-1}, which suggests that the distribution of M clusters more closely about μ than the distribution of X. The values of $E(M)$ and var (M) thus provide support for the procedure of estimating the population mean by the sample mean. However, a complete analysis of the procedure will involve the distribution of M, which depends on the distribution of X and the sample size. We return to this question in the next chapter.

The rth sample moment about the mean is denoted by m_r and defined by

$$m_r = \sum_{i=1}^{n} (x_i - m)^r/n \qquad (r = 2, 3, \ldots).$$

We omit the limits of summation in the rest of this section, since they remain the same. By analogy with the central moments of a probability distribution, we pay particular attention to

$$v = \Sigma(x_i - m)^2/n,$$

which is an observation on the variate

$$V = \Sigma(X_i - M)^2/n.$$

In discussing either of these quantities, we note that the identity

$$\Sigma(x_i - m)^2 \equiv \Sigma x_i^2 - 2m\Sigma x_i + nm^2$$

leads to two other expressions for $\Sigma(x_i - m)^2$, namely

$$\Sigma x_i^2 - nm^2 \quad \text{and} \quad \Sigma x_i^2 - (\Sigma x_i)^2/n.$$

From the first of the corresponding results for $\Sigma(X_i - M)^2$,

$$\begin{aligned} E\{\Sigma(X_i - M)^2\} &= E(\Sigma X_i^2) - nE(M^2) \\ &= n\{\text{var}\,(X) + \mu^2\} - n\{\text{var}\,(M) + \mu^2\} \\ &= n\{\text{var}\,(X) - \text{var}\,(M)\}. \end{aligned}$$

Inserting the value of var (M) derived above, we obtain

$$E(V) = (n-1)\sigma^2/n.$$

Thus $E(V) < \sigma^2$ and V is a biased estimator of σ^2.

The bias can be removed by introducing the statistic

$$s^2 = \Sigma(x_i - m)^2/(n-1),$$

corresponding to the variate

$$S^2 = \Sigma(X_i - M)^2/(n-1).$$

Both s^2 and S^2 are described as the **sample variance**. We now have

$$E(S^2) = E\{nV/(n-1)\} = \sigma^2,$$

so that S^2 is an unbiased estimator of σ^2. This result supports the use of the sample variance to estimate the population variance, but the distribution of S^2 is required before making a full assessment of the sample variance for estimation purposes.

The non-negative square root of s^2 is the **sample standard deviation** s. Since var (S) is non-negative,

$$E(S^2) - \{E(S)\}^2 \geqslant 0$$

and therefore

$$E(S) \leqslant \sigma.$$

In any practical situation, var $(S) > 0$ and S is then a biased estimator of σ. Similarly $S/\sqrt{n}$ is a biased estimator of the standard deviation of M. The corresponding estimate $s/\sqrt{n}$ is known as the **standard error** of the sample mean.

Other moments of M and S^2 are derived by arguments which can be lengthy but remain straightforward. For example, consider the evaluation of $\operatorname{cov}(M, S^2)$. The covariance of two variates is unaffected by change of origin. We therefore take $\mu = 0$, so that $\mu'_r = \mu_r$ $(r = 2, 3, \ldots)$ and

$$\operatorname{cov}(M, S^2) = E(MS^2).$$

The definitions of M and S^2 give

$$\begin{aligned} MS^2 &= \Sigma X_i\{\Sigma X_j^2 - (\Sigma X_j)^2/n\}/n(n-1) \\ &= \Sigma X_i \Sigma X_j^2/n(n-1) - (\Sigma X_i)^3/n^2(n-1). \end{aligned}$$

We collect the individual terms in the expansion of $\Sigma X_i\, \Sigma X_j^2$ according as $i = j$ or $i \neq j$, and obtain

$$\Sigma X_i \Sigma X_j^2 = \Sigma X_i^3 + \sum_{i \neq j} X_i X_j^2.$$

The expectation of the second term on the right is

$$E(\sum_{i \neq j} X_i X_j^2) = \sum_{i \neq j} E(X_i X_j^2),$$

and since $X_1, X_2, \ldots, X_n$ are mutually independent,

$$E(X_i X_j^2) = E(X_i)E(X_j^2).$$

Every such product vanishes because $\mu = 0$, and hence

$$E(\Sigma X_i \Sigma X_j^2) = E(\Sigma X_i^3) = n\mu_3.$$

Similarly, the individual items in the expansion of

$$(\Sigma X_i)^3 = (\Sigma X_i)(\Sigma X_j)(\Sigma X_k)$$

are collected according as three suffices are equal, two are equal, or all are different. Thus

$$(\Sigma X_i)^3 = \Sigma X_i^3 + 3 \sum_{i \neq j} X_i X_j^2 + \sum_{i, j, k \text{ different}} X_i X_j X_k.$$

The expectations of the second and third terms on the right are zero, whence

$$E\{(\Sigma X_i)^3\} = E(\Sigma X_i^3) = n\mu_3.$$

Inserting the values of $E(\Sigma X_i \Sigma X_j^2)$ and $E(\Sigma X_i^3)$ in $E(MS^2)$, we find that

$$E(MS^2) = \mu_3/(n-1) - \mu_3/n(n-1),$$

and consequently

$$\text{cov}\,(M, S^2) = \mu_3/n.$$

The same method of approach leads to

$$\text{var}\,(S^2) = (\mu_4 - \mu_2^2)/n + 2\mu_2^2/n(n-1).$$

Similar results can be obtained for a random sample of size n from the bivariate distribution of X and Y, when the sample pairs $(x_1, y_1), (x_2, y_2), \ldots, (x_n, y_n)$ are observations on $(X_1, Y_1), (X_2, Y_2), \ldots, (X_n, Y_n)$ respectively. Write $m_x = \Sigma x_i/n$ and $m_y = \Sigma y_i/n$. The **sample covariance** is

$$\Sigma(x_i - m_x)(y_i - m_y)/(n-1),$$

and it can be alternatively expressed as

$$(\Sigma x_i y_i - n m_x m_y)/(n-1) \quad \text{or} \quad \{\Sigma x_i y_i - (\Sigma x_i)(\Sigma y_i)/n\}/(n-1).$$

When evaluating the expectation of the sample covariance, we can take both $E(X) = 0$ and $E(Y) = 0$ because the covariance is unaffected by a change of origin in the observations. Consider

$$E\{\Sigma X_i Y_i - (\Sigma X_i)(\Sigma Y_i)/n\} = E(\Sigma X_i Y_i) - E(\Sigma\Sigma X_i Y_j)/n.$$

The expectations we require can be obtained from

$$E(X_i Y_j) = \begin{cases} \text{cov}\,(X, Y) & \text{if } i = j, \\ 0 & \text{if } i \neq j. \end{cases}$$

Inserting the appropriate values, we find that

$$E\{\Sigma X_i Y_i - (\Sigma X_i)(\Sigma Y_i)/n\} = (n-1)\,\text{cov}\,(X, Y).$$

Thus the sample covariance is an unbiased estimator of the population covariance.

§29. Calculation of sample mean and variance

The methods which follow are appropriate when computing facilities are limited. A linear transformation of the original measurements $x_1, x_2, \ldots, x_n$ results in a set of values $y_1, y_2, \ldots, y_n$ which are more convenient for numerical work. Suppose that

$$x_i = cy_i + d \qquad (i = 1, 2, \ldots, n).$$

Then d is called the **working origin** and c the **scale factor**. Denote the sample mean and variance of $\{x_i\}$ by m_x and s_x^2 respectively, and let the corresponding quantities for $\{y_i\}$ be m_y and s_y^2. Then

$$\Sigma x_i = c\Sigma y_i + nd,$$

whence

$$m_x = cm_y + d.$$

Again

$$\Sigma(x_i - m_x)^2 = c^2\Sigma(y_i - m_y)^2,$$

whence

$$s_x^2 = c^2 s_y^2.$$

Thus m_x and s_x^2 can be derived by simple formulae from m_y and s_y^2. In situations where the data consist of measurements $x_1, x_2, \ldots, x_k$ with frequencies $n_1, n_2, \ldots, n_k$ respectively, where $\Sigma n_j = n$, the same method applies. The sample mean and variance of the transformed values are now

$$m_y = \Sigma n_j y_j / n$$

and

$$s_y^2 = \Sigma n_j (y_j - m_y)^2 / (n-1),$$

where two other expressions for $\Sigma n_j(y_j - m_y)^2$ are

$$\Sigma n_j y_j^2 - n m_y^2 \quad \text{and} \quad \Sigma n_j y_j^2 - (\Sigma n_j y_j)^2/n.$$

Example 1. The following observations on the acceleration due to gravity are expressed in cm./sec.^{2}:

981·9 980·2 985·5 982·4 978·5

Here $n = 5$. Take $d = 980{\cdot}0$ and $c = 0{\cdot}1$. The values of $\{y_i\}$ are

19 2 55 24 −15

We find $\Sigma y_i = 85$ and $\Sigma y_i^2 = 4191$, whence

$$\Sigma(y_i - m_y)^2 = 4191 - 85^2/5 = 2746.$$

Consequently $m_y = 17,\ s_y^2 = 686{\cdot}5$

and finally $m_x = 981{\cdot}7,\ s_x^2 = 6{\cdot}865.$

The method may be contrasted with the direct evaluation of m_x and s_x^2 from the values $\{x_i\}$, which involves the manipulation of much larger numbers.

Example 2. Consider the frequency distribution of 688 measurements which is described in §**13** and displayed in Table 6 (p. 41). We assume that the values of x in each interval are all equal to the value at the centre of the interval e.g. the 65 values between 2·5 and 3·5 are all equal to 3. Take $d = 5$ and $c = 1$. The distinct values of y_j and the frequencies n_j with which they occur are as follows:

y_j	−4	−3	−2	−1	0	1	2	3	4	5
n_j	2	25	65	127	198	168	76	24	2	1

We find $\Sigma n_j y_j = 65$ and $\Sigma n_j y_j^2 = 1389$, whence

$$\Sigma n_j (y_j - m_y)^2 = 1389 - 65^2/688 = 1382{\cdot}86.$$

Consequently $$m_y = 0{\cdot}094,\ s_y^2 = 2{\cdot}013,$$

and finally $$m_x = 5{\cdot}094,\ s_x^2 = 2{\cdot}013.$$

The sample mean and standard deviation are 5·094 and 1·418 respectively, which are the values used to estimate μ and σ in §**13**.

§30. Order statistics

Suppose that a random sample of size n is taken from a continuous distribution. The sample values $x_1, x_2, \ldots, x_n$ are then distinct because the probability is zero that two or more of the corresponding variates are equal. Arrange $x_1, x_2, \ldots, x_n$ in order of magnitude and denote the smallest by x_{1n}, the next smallest by $x_{2n}, \ldots$, and the largest by x_{nn}. Thus

$$x_{1n} < x_{2n} < \ldots < x_{nn}.$$

The values $x_{1n}, x_{2n}, \ldots, x_{nn}$ are known as the **order statistics** of the sample, and various sample characteristics can be defined in terms of them. For instance, the **sample median** q is defined by

$$q = \begin{cases} x_{r+1,n} & (n = 2r+1), \\ \frac{1}{2}(x_{rn} + x_{r+1,n}) & (n = 2r). \end{cases}$$

Thus the sample median is the middle member of the ordered values when the sample size is odd, and the half-way point between the two middle values when the sample size is even. Another function of the order statistics is the **sample range** w, which is the difference between the largest and smallest values. Thus

$$w = x_{nn} - x_{1n}.$$

The sample mean and median are comparable, since each is a value which is representative of the sample observations. In the same way, the sample standard deviation and range both measure the dispersion of the sample values. The median and range are easily calculated and this recommends them for simple methods of analysing data.

We proceed to discuss the distribution theory of functions of the order statistics. Suppose that X is a variate with density function $f(x)$ and distribution function $F(x)$. Let $X_1, X_2, \ldots, X_n$ be mutually independent variates, each distributed like X. Arrange them in order of magnitude and denote the ordered variates by $X_{1n}, X_{2n}, \ldots, X_{nn}$, where

$$X_{1n} < X_{2n} < \ldots < X_{nn}.$$

The corresponding observations are the order statistics of a sample of size n from the distribution of X. A particular set of order statistics $x_{1n}, x_{2n}, \ldots, x_{nn}$ can arise from any of the $n!$ permutations of these values assigned to $X_1, X_2, \ldots, X_n$. Each permuted set of values has the same probability element, namely

$$f(x_{1n})\,f(x_{2n}) \ldots f(x_{nn})\,dx_{1n}\,dx_{2n} \ldots dx_{nn},$$

because $X_1, X_2, \ldots, X_n$ are mutually independent. Hence the joint density function of $X_{1n}, X_{2n}, \ldots, X_{nn}$ is

$$b(x_{1n}, x_{2n}, \ldots, x_{nn}) = n!\,f(x_{1n})\,f(x_{2n}) \ldots f(x_{nn}) \qquad (x_{1n} < x_{2n} < \ldots < x_{nn}).$$

The distribution of any function of the order statistics can now be derived by a transformation of the variables. However, we can also argue directly from the original variates, as examples will show.

Consider first the distribution of X_{kn}, where $1 \leqslant k \leqslant n$. The probability that i of the variates $X_1, X_2, \ldots, X_n$ are less than y and $n-i$ exceed y is

$$\binom{n}{i} \{F(y)\}^i \{1-F(y)\}^{n-i}.$$

But $X_{kn} < y$ if and only if $i \geqslant k$. Therefore

$$p(X_{kn} < y) = \sum_{i=k}^{n} \binom{n}{i} \{F(y)\}^i \{1-F(y)\}^{n-i}.$$

This is the distribution function of X_{kn}, and the density function is

$$c_{kn}(y) = \sum_{i=k}^{n} \frac{d}{dy}\left[\binom{n}{i} \{F(y)\}^i \{1-F(y)\}^{n-i}\right].$$

The derivatives are $n\{F(y)\}^{n-1}f(y)$ for $i = n$ and

$$n\binom{n-1}{i-1}\{F(y)\}^{i-1}\{1-F(y)\}^{n-i}f(y) - n\binom{n-1}{i}\{F(y)\}^{i}\{1-F(y)\}^{n-i-1}f(y)$$

for $i < n$. When the derivatives are added for $i = k, k+1, \ldots, n$, all the individual terms cancel out except the first of the pair for $i = k$. Hence the density function of X_{kn} is

$$c_{kn}(y) = n\binom{n-1}{k-1}\{F(y)\}^{k-1}\{1-F(y)\}^{n-k}f(y).$$

The expected value of the kth order statistic in a sample of size n is given by

$$E(X_{kn}) = \int_{-\infty}^{\infty} y c_{kn}(y)\, dy.$$

Suppose that X has a symmetrical distribution with mean μ, so that

$$F(y) = 1-F(2\mu-y) \quad \text{and} \quad f(y) = f(2\mu-y).$$

Substitute $y = 2\mu-x$ in the integral for $E(X_{kn})$. Then

$$E(X_{kn})$$
$$= \int_{-\infty}^{\infty} (2\mu-x)n\binom{n-1}{k-1}\{(F(2\mu-x)\}^{k-1}\{1-F(2\mu-x)\}^{n-k}f(2\mu-x)\,dx$$
$$= \int_{-\infty}^{\infty} (2\mu-x)n\binom{n-1}{n-k}\{1-F(x)\}^{k-1}\{F(x)\}^{n-k}f(x)\,dx.$$

The integrand is $(2\mu-x)c_{n-k+1,n}(x)$, and hence

$$E(X_{kn}) = 2\mu - E(X_{n-k+1,n}).$$

This relates the expected value of the kth order statistic in ascending order of magnitude to that of the kth in descending order. An immediate consequence is that $\frac{1}{2}(X_{kn}+X_{n-k+1,n})$ is an unbiased estimator of μ, a result which is intuitively acceptable on grounds of symmetry. When $n = 2r+1$ and $k = r+1$, this estimator is the sample median.

Example. In the previous section we gave the following observations on the acceleration due to gravity expressed in cm./sec.^{2}:

$$981{\cdot}9 \qquad 980{\cdot}2 \qquad 985{\cdot}5 \qquad 982{\cdot}4 \qquad 978{\cdot}5$$

On the assumption that the observations form a random sample from an arbitrary symmetrical distribution, we can estimate μ by the sample median, 981·9. This is closer to the sample mean, 981·7, than would usually occur with σ about 2 or 3. Other estimates are

$$\tfrac{1}{2}(978{\cdot}5+985{\cdot}5) = 982{\cdot}0 \quad \text{and} \quad \tfrac{1}{2}(980{\cdot}2+982{\cdot}4) = 981{\cdot}3.$$

This multiplicity of estimates is typical of statistical problems and indicates the need for methods of discriminating between the possibilities based on the relevant probability distributions. We return to this problem in the next chapter.

Consider the interval between any two order statistics of an arbitrary continuous distribution, and denote the length of this interval by the variate L. A simple formula for the expectation of L is established as follows. Let $I(x)$ be a variate which takes the value 1 when the interval contains the point x, and is otherwise 0. Let $p(x)$ be the probability that the interval contains x. Then the probability distribution of $I(x)$ is discrete, with probabilities $p(x)$ at 1 and $1-p(x)$ at 0. Hence

$$E\{I(x)\} = p(x).$$

The length of the interval can be expressed in the form

$$L = \int_{-\infty}^{\infty} I(x)\,dx$$

and therefore

$$E(L) = \int_{-\infty}^{\infty} p(x)\,dx.$$

We turn now to the sample range and write W for $X_{nn}-X_{1n}$. The range will fail to contain a point x if $X_1, X_2, \ldots, X_n$ are all less than x, or if they all exceed x. Since the corresponding probabilities are $\{F(x)\}^n$ and $\{1-F(x)\}^n$ respectively, we have

$$p(x) = 1-\{F(x)\}^n-\{1-F(x)\}^n.$$

An application of the formula just derived shows that

$$E(W) = \int_{-\infty}^{\infty} [1-\{F(x)\}^n-\{1-F(x)\}^n]\,dx.$$

The distribution of the range can be obtained by a similar interpretation of the probabilities involved. Given x and y with $x < y$, the probability that $X_1, X_2, \ldots, X_n$ all lie in (x, y) is $\{F(y)-F(x)\}^n$. Thus

$$p(X_{nn} < y, X_{1n} > x) = \{F(y)-F(x)\}^n.$$

The joint density function of X_{1n} and X_{nn} is therefore

$$-\frac{\partial^2}{\partial x \partial y} p(X_{nn} < y, X_{1n} > x) = n(n-1)\{F(y)-F(x)\}^{n-2}f(y)f(x).$$

Make the transformation $w = y-x$, $u = x$. The density function of W is

$$c_n(w) = n(n-1)\int \{F(u+w)-F(u)\}^{n-2}f(u+w)f(u)\,du.$$

Care is necessary in distinguishing between the various forms taken by F when f vanishes for some values of the argument.

EXERCISES 6

1. X is a continuous variate with density function $f(x)$ and distribution function $F(x)$. Define $U = F(X)$, which is known as the **probability integral transformation**. Show that U is distributed uniformly on (0, 1). In §27, X was obtained from U by solving $U = F(X)$.

2. X is a discrete variate such that

$$p(X = x_j) = f_j \qquad (j = 0, 1, 2, \ldots),$$

where $x_j < x_{j+1}$ for all j. The distribution function is $F(x)$. Find the distribution of $F(X)$, and show that

$$E\{F(X)\} = \tfrac{1}{2}(1+\Sigma f_j^2).$$

3. U_1 and U_2 are independently and uniformly distributed on (0, 1). Show that the variates

$$X_1 = (-2\log U_1)^{\frac{1}{2}}\cos(2\pi U_2)$$

and

$$X_2 = (-2\log U_1)^{\frac{1}{2}}\sin(2\pi U_2)$$

are independently and normally distributed with zero mean and unit variance. This transformation, which is due to Box and Muller, is used to convert pairs of pseudo-uniform variates generated by a computer into pairs of pseudo-normal variates for simulation purposes.

4. A sample of size n is taken from the distribution of X and the sample mean is M. Denote the central moments of X by $\{\mu_r\}$ and those of M by $\{\nu_r\}$. Show that the moment generating functions of M and X are related by

$$J_m(z) = \{J_x(z/n)\}^n.$$

Hence or otherwise prove that $\nu_3 = \mu_3/n^2$, and

$$(\nu_4/\nu_2^2)-3 = \{(\mu_4/\mu_2^2)-3\}/n.$$

5. A sample of size n consists of observations on $X_1, X_2, \ldots, X_n$. The mean square successive difference is denoted by R^2 and defined by

$$R^2 = \sum_{i=1}^{n-1} (X_i - X_{i+1})^2/2(n-1).$$

Show that R^2 is an unbiased estimator of σ^2 and that

$$\operatorname{var}(R^2) = \mu_4/(n-1)-(\mu_4-\mu_2^2)/2(n-1)^2.$$

Verify that

$$\operatorname{var}(S^2) = (\mu_4-\mu_2^2)/n+2\mu_2^2/n(n-1).$$

Compare R^2 and S^2 as estimators of σ^2 in samples from a normal distribution.

6. A sample of size $n = 2r+1$ is taken from the distribution of a continuous variate X, and the sample values are $x_1, x_2, \ldots, x_n$. Show that the function of x defined by $\Sigma \mid x_i - x \mid$ attains its minimum value when x is the sample median $x_{r+1,n}$.

Prove that the median of $X_{r+1,n}$ is also the median of X.

7. X has a uniform distribution on the interval $(0, 1)$. Find the density function of $X_{a,\,a+b-1}$, and hence show that

$$\int_0^1 x^{a-1}(1-x)^{b-1}\,dx = (a-1)!\,(b-1)!\,/\,(a+b-1)!.$$

This integral is known as a beta function and usually denoted by $B(a, b)$. Show also that

$$E(X_{a,\,a+b-1}) = a/(a+b)$$

and

$$\operatorname{var}(X_{a,\,a+b-1}) = ab/(a+b)^2(a+b+1).$$

8. A sample of size n is taken from the uniform distribution on $(0, 1)$. Suppose that $1 \leqslant j < k \leqslant n$. Show that the joint density function of X_{jn} and X_{kn} is

$$n!\, x_{jn}^{j-1}(x_{kn}-x_{jn})^{k-j-1}(1-x_{kn})^{n-k}/(j-1)!\,(k-j-1)!\,(n-k)!\,.$$

Hence prove that

$$\operatorname{cov}(X_{jn}, X_{kn}) = j(n+1-k)/(n+1)^2(n+2) \qquad (j \leqslant k).$$

9. Prove that the distribution function of the range is given by

$$C_n(w) = n \int \{F(u+w)-F(u)\}^{n-1} f(u)\, du.$$

When X is distributed uniformly on $(0, 1)$, deduce that

$$C_n(w) = \begin{cases} 0 & (w \leqslant 0), \\ w^{n-1}\{n-(n-1)w\} & (0 < w < 1), \\ 1 & (w \geqslant 1). \end{cases}$$

Verify that the mean range derived from this sampling distribution checks with the value obtained directly from the formula for $E(W)$.

10. X has an exponential distribution with parameter λ. A transformation from the order statistics of a sample of size n to $y_1, y_2, \ldots, y_n$ is defined by

$$x_{kn} = y_1+y_2+\ldots+y_k \qquad (k = 1, 2, \ldots, n).$$

Prove that $Y_1, Y_2, \ldots, Y_n$ are mutually independent and that Y_i has an exponential distribution with parameter $(n-i+1)\lambda$. Deduce that

$$E(X_{kn}) = \frac{1}{\lambda}\sum_{i=1}^{k}\frac{1}{n-i+1} \quad \text{and} \quad \operatorname{var}(X_{kn}) = \frac{1}{\lambda^2}\sum_{i=1}^{k}\frac{1}{(n-i+1)^2}.$$

CHAPTER 7
STATISTICAL INFERENCE

§31. Statistical inference

An analysis of experimental data must necessarily be based on a mathematical structure which fully describes how the data could be generated and which is known as a **statistical model**. The construction of appropriate models calls for wide experience of statistical practice and we therefore confine attention to models which are already specified. A simple class of statistical models arises from the concept of a **family** of distributions.

Suppose that the distribution of a variate X has a known mathematical form, but the value of a parameter θ is required in order to complete the specification. When θ varies over the range of its possible values, a family of distributions is obtained. We say that the family is **indexed** by θ because the value of θ identifies a particular distribution. For example, X has a Poisson distribution with mean θ, or X has a normal distribution with mean θ and variance 1. When the distribution of X has a known mathematical form but the values of two parameters θ and β are required to complete the specification, we similarly obtain a family of distributions indexed by θ and β. For example, X has a normal distribution with mean θ and variance β.

This chapter is concerned with statistical models of the following type. The data consist of a random sample from a distribution which belongs to a known family indexed by θ. We require to extract from the sample data as much information as possible about the value of θ which identifies the distribution sampled. There are various methods of presenting the information. We can assign a number to θ, or construct an interval in which θ may be supposed to lie, or we can examine the agreement between the data and a hypothetical value of θ. Every such procedure represents an inference about the population from the sample. When the same procedure is used in repeated samples from the same population, the inferences about θ change at random. The procedure must therefore be judged by the statistical properties of the resulting inferences, and we therefore describe this type of argument from sample to population as **statistical inference**.

§32. Likelihood

We shall employ a notation for probability distributions which can be interpreted for either discrete or continuous variates as required. Suppose that the distribution of X belongs to a family of distributions indexed by a parameter θ. When X is a discrete variate, we write

$$p(X = x) = f(x; \theta),$$

where x ranges over an enumerable set of values. When X is a continuous variate, $f(x; \theta)$ denotes the probability density function. In each case, $f(x; \theta)$ is regarded as a function of x for a fixed value of θ. Similarly, when the distribution of X belongs to a family indexed by two parameters θ and β, $f(x; \theta, \beta)$ represents the probability distribution or the probability density function according as X is a discrete or continuous variate respectively.

We now introduce a basic concept of statistical inference. Suppose that $x_1, x_2, \ldots, x_n$ are the observations in a random sample of size n, and let $X_1, X_2, \ldots, X_n$ be the corresponding variates. Thus $X_1, X_2, \ldots, X_n$ are mutually independent variates each distributed like X. The **likelihood** of $x_1, x_2, \ldots, x_n$ is defined as follows. If X is discrete, then the likelihood is the probability that $X_1, X_2, \ldots, X_n$ take the values $x_1, x_2, \ldots, x_n$ respectively, and if X is continuous, then the likelihood is the joint density function of $X_1, X_2, \ldots, X_n$ evaluated at $x_1, x_2, \ldots, x_n$ respectively. When X belongs to a family of distributions indexed by θ, the likelihood of $x_1, x_2, \ldots, x_n$ is denoted by $l(\theta; x_1, x_2, \ldots, x_n)$. Thus

$$l(\theta; x_1, x_2, \ldots, x_n) = f(x_1; \theta) f(x_2; \theta) \ldots f(x_n; \theta).$$

For any given set of sample values, this is a function of θ known as the **likelihood function**. Similarly, when X belongs to a family of distributions indexed by θ and β, the likelihood of $x_1, x_2, \ldots, x_n$ is denoted by $l(\theta, \beta; x_1, x_2, \ldots, x_n)$. Thus

$$l(\theta, \beta; x_1, x_2, \ldots, x_n) = f(x_1; \theta, \beta) f(x_2; \theta, \beta) \ldots f(x_n; \theta, \beta).$$

The likelihood is now a function of θ and β.

Example 1. X has a Poisson distribution with mean θ, so that

$$f(x; \theta) = \theta^x e^{-\theta}/x! \qquad (x = 0, 1, 2, \ldots).$$

The likelihood function is

$$l(\theta; x_1, x_2, \ldots, x_n) = \Pi(\theta^{x_i} e^{-\theta}/x_i!) = \theta^{\Sigma x_i} e^{-n\theta}/\Pi x_i!,$$

where the products and summations are taken over $i = 1, 2, \ldots, n$. Alternatively, we can argue as follows. Each of $x_1, x_2, \ldots, x_n$ is a non-negative integer, and hence the sample values consist of v with frequency n_v, where $v = 0, 1, 2, \ldots$, and $\Sigma n_v = n$. Thus

$$l(\theta; x_1, x_2, \ldots, x_n) = \theta^{\Sigma v n_v} e^{-n\theta}/\Pi(v!)^{n_v}.$$

Example 2. X has a normal distribution with mean θ and variance 1, so that

$$f(x; \theta) = (2\pi)^{-\frac{1}{2}} \exp\{-\tfrac{1}{2}(x-\theta)^2\}.$$

The likelihood function is

$$l(\theta; x_1, x_2, \ldots, x_n) = (2\pi)^{-\frac{1}{2}n} \exp\{-\tfrac{1}{2}\Sigma(x_i-\theta)^2\}.$$

Example 3. X has a normal distribution with mean θ and variance β, so that

$$f(x; \theta, \beta) = (2\pi\beta)^{-\frac{1}{2}} \exp\{-(x-\theta)^2/2\beta\}.$$

The likelihood function is

$$l(\theta, \beta; x_1, x_2, \ldots, x_n) = (2\pi\beta)^{-\frac{1}{2}n} \exp\{-\Sigma(x_i-\theta)^2/2\beta\}.$$

Throughout the remainder of this section, we make the following assumptions.

(*i*) The distribution of X belongs to a family of distributions indexed by a parameter θ.

(*ii*) The possible values of X are independent of θ.

(*iii*) The first two derivatives of the likelihood function with respect to θ exist for all possible values of θ and $x_1, x_2, \ldots, x_n$.

Denote the likelihood $l(\theta; x_1, x_2, \ldots, x_n)$ by l and the corresponding variate $l(\theta; X_1, X_2, \ldots, X_n)$ by L. We shall prove that

$$E(\partial \log L/\partial\theta) = 0 \quad \text{and} \quad \text{var}(\partial \log L/\partial\theta) = -E(\partial^2 \log L/\partial\theta^2).$$

In the proof which follows, X is assumed to be a continuous variate. The proof for a discrete variate is identical apart from minor changes of notation. Since the total probability in the outcome space is unity, we have

$$\int\int \ldots \int l \, dx_1 \, dx_2 \ldots dx_n = 1.$$

This is true for all θ and the limits of integration are independent of θ. Differentiation with respect to θ yields

$$\int\int \ldots \int (\partial l/\partial\theta) \, dx_1 \, dx_2 \ldots dx_n = 0.$$

which can be written in the form

$$\int\int \ldots \int (\partial \log l/\partial\theta) l \, dx_1 \, dx_2 \ldots dx_n = 0.$$

This expresses the fact that

$$E(\partial \log L/\partial\theta) = 0.$$

For the second part of the proof, we require

$$\frac{\partial^2 \log l}{\partial\theta^2} = \frac{\partial}{\partial\theta}\left(\frac{1}{l}\frac{\partial l}{\partial\theta}\right) = \frac{1}{l}\frac{\partial^2 l}{\partial\theta^2} - \frac{1}{l^2}\left(\frac{\partial l}{\partial\theta}\right)^2.$$

Multiply by l and integrate over $x_1, x_2, \ldots, x_n$ whence

$$E\left(\frac{\partial^2 \log L}{\partial\theta^2}\right) = \int\int \ldots \int \left(\frac{\partial^2 l}{\partial\theta^2}\right) dx_1 \, dx_2 \ldots dx_n - E\left(\frac{1}{L}\frac{\partial L}{\partial\theta}\right)^2.$$

The integral is the second derivative of the total probability and therefore zero. Consequently

$$E(\partial^2 \log L/\partial\theta^2) = -E(\partial \log L/\partial\theta)^2.$$

However, $E(\partial \log L/\partial\theta) = 0$, and thus

$$\text{var}\,(\partial \log L/\partial\theta) = -E(\partial^2 \log L/\partial\theta^2),$$

which completes the proof. We note that

$$\log L = \Sigma \log f(X_i; \theta),$$

where each of the variates $f(X_1; \theta), f(X_2; \theta), \ldots, f(X_n; 0)$ is distributed like $f(X; \theta)$. Hence

$$\text{var}\,(\partial \log L/\partial\theta) = -n\, E\{\partial^2 \log f(X; \theta)/\partial\theta^2\}.$$

§33. Point estimation

A random sample is taken from a distribution which belongs to a known family indexed by a parameter θ. We distinguish between the procedure of assigning a number to θ, and the procedure of constructing an interval in which θ may be supposed to lie, by referring to the first as **point** estimation and the second as **interval** estimation. This section is solely concerned with point estimation and we can therefore omit the qualification 'point' without ambiguity. Suppose that $t(x_1, x_2, \ldots, x_n)$ is an arbitrary function of the sample observations $x_1, x_2, \ldots, x_n$ and denote the corresponding variate $t(X_1, X_2, \ldots, X_n)$ by T. Thus $t(x_1, x_2, \ldots, x_n)$ is an estimate of θ and T is the corresponding estimator.

The question naturally arises of which function t to choose. An estimation procedure is judged by its statistical properties and hence the principal factor which affects the choice of t is the probability distribution of the variate T. The comparison of different estimators is most easily accomplished by concentrating attention on the basic properties of a variate, such as measures of location and dispersion. We recall that an estimator is said to be unbiased if $E(T) = \theta$. Otherwise T is biased, and the bias is measured by $E(T)-\theta$. For example, the sample mean M is an unbiased estimator of the population mean μ, whereas $\Sigma(X_i-M)^2/n$ is a biased estimator of the population variance σ^2, with a bias of $-\sigma^2/n$. A zero bias is desirable because we prefer to exclude estimators which are generally too high or too low. However, the example in §**30** shows that there are usually many unbiased estimators, and we now face the task of choosing between them.

A method of discriminating between several unbiased estimators is to select the one for which $E\,|\,T-\theta\,|^r$ is least, where r is a specified positive integer. This can be justified on the following grounds. First, $E\,|\,T-\theta\,|^r$ measures the amount by which the distribution of T is dispersed about θ. Secondly, the loss incurred by taking T to be the actual value of θ can reasonably have the form $k\,|\,T-\theta\,|^r$ where k is a positive constant, and we prefer the estimator which minimizes the expected loss $kE\,|\,T-\theta\,|^r$. The convenient mathematical properties of the variance recommend $r = 2$ and lead to the requirement that T has minimum variance in the class of all unbiased estimators.

We proceed to derive a lower bound for the variance of an unbiased estimator, and determine the conditions under which the lower bound is attained. Suppose that X is a continuous variate and make the same assumptions about its distribution as those required to find the mean and variance of $\partial \log L/\partial\theta$. Let T be an unbiased estimator of θ and denote $t(x_1, x_2, \ldots, x_n)$ by t. Then

$$\int\int \ldots \int t\, l\, dx_1\, dx_2 \ldots dx_n = \theta$$

uniformly in θ. Differentiation with respect to θ yields

$$\int\int \ldots \int t(\partial l/\partial\theta)\, dx_1\, dx_2 \ldots dx_n = 1,$$

which can be rewritten as

$$\int\int \ldots \int t(\partial \log l/\partial\theta) l\, dx_1\, dx_2 \ldots dx_n = 1.$$

In terms of variates,

$$E(T\, \partial \log L/\partial\theta) = 1.$$

Now $E(\partial \log L/\partial\theta) = 0$, and hence

$$\text{cov}\,(T, \partial \log L/\partial\theta) = 1.$$

Consider the correlation coefficient of T and $\partial \log L/\partial\theta$.

$$\begin{aligned}\{\text{corr}\,(T, \partial \log L/\partial\theta)\}^2 & \\ &= \{\text{cov}\,(T, \partial \log L/\partial\theta)\}^2/\text{var}\,(T)\,\text{var}\,(\partial \log L/\partial\theta) \\ &= \{\text{var}\,(T)\,\text{var}\,(\partial \log L/\partial\theta)\}^{-1}.\end{aligned}$$

However, $\{\text{corr}\,(T, \partial \log L/\partial\theta)\}^2 \leqslant 1$, and therefore

$$\text{var}\,(T) \geqslant \{\text{var}\, \partial \log L/\partial\theta)\}^{-1}.$$

Substituting for var $(\partial \log L/\partial\theta)$, we have

$$\text{var}\,(T) \geqslant [-nE\{\partial^2 \log f(X;\theta)/\partial\theta^2\}]^{-1}.$$

This is known as the Cramér-Rao inequality, and the expression on the right gives a lower bound to the variance of an unbiased estimator of θ. The same result holds when X is a discrete variate and the proof follows identical lines.

Inspection of the proof shows that var (T) attains its lower bound when

$$\{\text{corr}\,(T, \partial \log L/\partial\theta)\}^2 = 1.$$

According to §**20**, the condition corr $(X, Y) = \pm 1$ implies that

$$(X-\mu_x)/\sigma_x = \pm(Y-\mu_y)/\sigma_y$$

respectively. Thus the lower bound of var (T) is attained when $\partial \log L/\partial\theta$ can be expressed in the form

$$\partial \log L/\partial\theta = a(\theta)(T-\theta),$$

where $\quad a^2(\theta) = \text{var}\,(\partial \log L/\partial\theta)/\text{var}\,(T) = \{\text{var}\,(T)\}^{-2}.$

We conclude that if $\partial \log L/\partial\theta$ has the form required for var (T) to attain its lower bound, then the value of the lower bound is $|a(\theta)|^{-1}$.

An unbiased estimator U with a variance equal to the lower bound of var (T) is known as a **minimum variance bound** estimator. The ratio var (U)/var (T) lies in $[0, 1]$ and is described as the **relative efficiency** of T as an estimator of θ.

Example 1. A sample of size n is taken from a Poisson distribution with mean θ. The likelihood function is

$$l(\theta; x_1, x_2, \ldots, x_n) = \theta^{\Sigma x_i} e^{-n\theta}/\Pi x_i!\,.$$

Consequently $$\log L = \Sigma X_i \log\theta - n\theta - \Sigma \log X_i!\,,$$

and $$\partial \log L/\partial\theta = \Sigma X_i/\theta - n = n(M-\theta)/\theta,$$

where M is the sample mean $\Sigma X_i/n$. Thus M is a minimum variance bound estimator of θ and var $(M) = \theta/n$.

Example 2. A sample of size n is taken from a normal distribution with mean θ and variance 1. The likelihood function is

$$l(\theta; x_1, x_2, \ldots, x_n) = (2\pi)^{-\frac{1}{2}n} \exp\{-\tfrac{1}{2}\Sigma(x_i-\theta)^2\}.$$

Consequently $$\log L = -\tfrac{1}{2}n \log(2\pi) - \tfrac{1}{2}\Sigma(X_i-\theta)^2,$$

and $$\partial \log L/\partial\theta = \Sigma(X_i-\theta) = n(M-\theta),$$

where M is the sample mean. Thus M is a minimum variance bound estimator of θ and var $(M) = 1/n$.

Example 3. A distribution which arises in high-energy physics has the density function

$$f(x;\theta) = \tfrac{1}{2}(1+\theta x) \qquad (-1 \leqslant x \leqslant 1,\ -1 \leqslant \theta \leqslant 1).$$

The mean and variance are calculated by straightforward integration. We find that

$$E(X) = \theta/3 \quad \text{and} \quad \text{var}\,(X) = (3-\theta^2)/9.$$

The likelihood function for a sample of size n is

$$l(\theta; x_1, x_2, \ldots, x_n) = 2^{-n}\Pi(1+\theta x_i).$$

Consequently $$\log L = -n\log 2 + \Sigma \log(1+\theta X_i),$$

and $$\partial \log L/\partial\theta = \Sigma\{X_i/(1+\theta X_i)\}.$$

The derivative cannot be expressed in the form required for a minimum variance bound estimator. We determine the lower bound of var (T) as follows.

$$\partial^2 \log f(X;\theta)/\partial\theta^2 = -X^2/(1+\theta X)^2,$$

and therefore

$$E\{\partial^2 \log f(X;\theta)/\partial\theta^2\} = -\tfrac{1}{2}\int_{-1}^{1} \{x^2/(1+\theta x)\}\,dx.$$

The integral is evaluated by substituting $y = 1+\theta x$, and we finally obtain

$$-1/n\,E\{\partial^2 \log f(X;\theta)/\partial\theta^2\} = 2\theta^3/n\left\{\log\left(\frac{1+\theta}{1-\theta}\right)-2\theta\right\}.$$

Consider the estimator $3M$. Its mean and variance are given by

$$E(3M) = 3E(X) = \theta,$$

and $$\operatorname{var}(3M) = 9\operatorname{var}(X)/n = (3-\theta^2)/n.$$

Thus $3M$ is an unbiased estimator of θ with a relative efficiency equal to

$$2\theta^3/(3-\theta^2)\left\{\log\left(\frac{1+\theta}{1-\theta}\right)-2\theta\right\} = (1+4\theta^2/15+8\theta^2/35+\ldots)^{-1}.$$

This is 1 when $\theta = 0$, 0·97 when $\theta = \pm 1/3$, 0·84 when $\theta = \pm 2/3$, and 0·35 when $\theta = \pm 49/50$.

§34. Sufficient statistics

Consider again the question of which function t should be chosen when making inferences about θ. Suppose that the likelihood of a sample of size n can be factorized in the form

$$l(\theta; x_1, x_2, \ldots, x_n) = g(t;\theta)h(x_1, x_2, \ldots, x_n).$$

When this condition holds, we say that t is a **sufficient statistic**, and that T is sufficient for θ. Let $j(t)$ be any function of t with a unique inverse $t(j)$. The factorization of the likelihood becomes

$$l(\theta; x_1, x_2, \ldots, x_n) = g\{t(j);\theta\}h(x_1, x_2, \ldots, x_n).$$

Thus any such function j is also a sufficient statistic.

Suppose that T is sufficient for θ. Then the joint distribution of any statistics other than t, conditional on a specified value of T, is independent of θ. This is proved for continuous variates as follows. Consider a transformation from the sample values $x_1, x_2, \ldots, x_n$ to t and $(n-1)$ other statistics, say $u_1, u_2, \ldots, u_{n-1}$. Denote the Jacobian by J and let $k(t, u_1, u_2, \ldots, u_{n-1})$ be the function into which $h(x_1, x_2, \ldots, x_n)$ is transformed. The joint density function of $T, U_1, U_2, \ldots, U_{n-1}$ is $g(t;\theta)kJ$, and the density function of T is

$$g(t;\theta)\int\int\ldots\int kJ\,du_1\,du_2\ldots du_{n-1}.$$

Hence the joint density function of $U_1, U_2, \ldots, U_{n-1}$ conditional on $T = t$ is

$$kJ/\int\int\ldots\int kJ\,du_1\,du_2\ldots du_{n-1},$$

which is independent of θ. Thus t is a sufficient statistic in the sense that, when the value of T is given, all other statistics are irrelevant as far as inferences about θ are concerned.

Example 1. X has a normal distribution with mean θ and variance 1. The likelihood function is

$$l(\theta; x_1, x_2, \ldots, x_n) = (2\pi)^{-\frac{1}{2}n}\exp\{-\tfrac{1}{2}n(m-\theta)^2-\tfrac{1}{2}\Sigma(x_i-m)^2\},$$

where m denotes $\Sigma x_i/n$. Hence $\Sigma X_i/n$ is sufficient for θ, as also is ΣX_i.

Example 2. X has the density function

$$f(x) = \tfrac{1}{2}(1+\theta x) \qquad (-1 \leqslant x \leqslant 1,\ -1 \leqslant \theta \leqslant 1).$$

The likelihood function cannot be expressed in the form required for a sufficient statistic.

Suppose that T is sufficient for θ, and that U is an unbiased estimator of θ. The quantity $E(U \mid T)$ is a function of T and is therefore also sufficient for θ. Under a condition stated below, $E(U \mid T)$ is the unique unbiased estimator of θ which has minimum variance in the class of all unbiased estimators. This result, which is due to Rao and Blackwell, is proved as follows. First, a reference to §**20** shows that

$$E\{E(U \mid T)\} = E(U).$$

Consequently $E(U \mid T)$ is an unbiased estimator of θ. Secondly, a reference to exercise 9 in chapter 4 shows that

$$\text{var}(U) = E\{\text{var}(U \mid T)\}+\text{var}\{E(U \mid T)\}.$$

The first term on the right is non-negative, whence

$$\text{var}\{E(U \mid T)\} \leqslant \text{var}(U).$$

We now reach the final stage of the proof. Denote the density function of T by $c(t; \theta)$. The family of distributions typified by $c(t; \theta)$ is said to be **complete** if the only function $b(t)$ which satisfies the equation

$$\int b(t)c(t; \theta)\, dt = 0$$

for all possible values of θ is $b(t) \equiv 0$. Suppose that U_1 and U_2 are any two unbiased estimators of θ. Then

$$E\{E(U_1 \mid T)\} = E\{E(U_2 \mid T)\},$$

or equivalently

$$\int \{E(U_1 \mid t) - E(U_2 \mid t)\} c(t; \theta)\, dt = 0$$

If the family of distributions $c(t; \theta)$ is complete, then the only solution of this equation valid for all θ is

$$E(U_1 \mid t) = E(U_2 \mid t),$$

so that $E(U \mid T)$ does not depend on the choice of U. Therefore, when completeness obtains, $E(U \mid T)$ is the unique unbiased estimator of θ with minimum variance.

Several of the families of distributions which are useful in statistical practice lead to distributions of T which are complete, but the question of whether or not this property holds must be established by special arguments mounted for individual cases.

Example 3. X has a Poisson distribution with mean θ. We require to estimate θ^2. The likelihood function is

$$l(\theta; x_1, x_2, \ldots, x_n) = \theta^{\Sigma x_i} e^{-n\theta} / \Pi x_i! \,.$$

Hence ΣX_i is sufficient for θ. The probability generating function of ΣX_i is $\exp\{n\theta(s-1)\}$. Therefore ΣX_i has a Poisson distribution with mean $n\theta$, say λ. The only function $b(k)$ which satisfies

$$\sum_{k=0}^{\infty} b(k) \lambda^k e^{-\lambda}/k! = 0$$

for all $\lambda > 0$ is $b(k) \equiv 0$. This shows that the family of Poisson distributions is complete. Denote the sample mean by M. Then $E(M) = \theta$ and $\text{var}(M) = \theta/n$, whence

$$E(M^2 - M/n) = \theta^2.$$

Thus $M^2 - M/n$, which is a function of ΣX_i, is also an unbiased estimator of θ^2. We conclude that $M^2 - M/n$ is the unique unbiased estimator of θ^2 with minimum variance.

§35. Principle of maximum likelihood

We now introduce a general method of point estimation. Suppose first that a random sample is taken from a distribution which belongs to a known family indexed by a parameter θ. Examples 1 and 2 in §**33** show that if T is a minimum variance bound estimator of θ, then T is obtained by solving the equation

$$\partial \log L/\partial\theta = 0.$$

Furthermore, $\text{var}(T) \geqslant 0$, whence $E(\partial^2 \log L/\partial\theta^2) \leqslant 0$. Consider the procedure of estimating θ by the function $\hat{\theta}(x_1, x_2, \ldots, x_n)$ which maximizes $\log l$ with respect to variation in θ. This procedure leads to a minimum variance bound estimator of θ when one exists, because if $\log l$ is differentiable, then at a maximum

$$\partial \log l/\partial\theta = 0 \quad \text{and} \quad \partial^2 \log l/\partial\theta^2 \leqslant 0.$$

The likelihood function l is maximized at the same value of θ. We therefore say that the estimation procedure described is obtained from the **principle of maximum likelihood**. The function $\hat{\theta}(x_1, x_2, \ldots, x_n)$ which maximizes $\log l$ is the maximum likelihood estimate of θ and $\hat{\theta}(X_1, X_2, \ldots, X_n)$ is the maximum likelihood estimator. Read $\hat{\theta}$ as 'θ hat'. If a sufficient statistic exists, then

$$\partial \log l/\partial\theta = \partial \log g(t; \theta)/\partial\theta.$$

Thus $\hat{\theta}$ is a function of T and is sufficient for θ. Further reasons for the use of maximum likelihood estimators are given in chapter 10.

Example 1. A sample of size n is taken from a Poisson distribution with mean θ. In example 1 of §**33**, we showed that

$$\partial \log l/\partial\theta = n(m-\theta)/\theta,$$

where m is the sample mean. Thus m is the maximum likelihood estimate of θ.

Example 2. A sample of size n is taken from a normal distribution with mean θ and variance 1. In example 2 of §**33**, we showed that

$$\partial \log l/\partial\theta = n(m-\theta).$$

Thus m is the maximum likelihood estimate of θ.

In many problems, the maximum likelihood estimate is not an

explicit function of $x_1, x_2, \ldots, x_n$. However, for any given $x_1, x_2, \ldots, x_n$, the estimate can usually be computed by an application of the Newton-Raphson method of solving equations. We proceed as follows. Write $g(\theta)$ for $\partial \log l/\partial\theta$ and let $\hat{\theta}$ be the maximum likelihood estimate, so that $g(\hat{\theta}) = 0$. Suppose that θ_1 is an approximation to $\hat{\theta}$, in which case

$$g(\hat{\theta}) \simeq g(\theta_1)+(\hat{\theta}-\theta_1)g'(\theta_1).$$

Define θ_2 as the solution of

$$g(\theta_1)+(\theta_2-\theta_1)g'(\theta_1) = 0,$$

whence
$$\theta_2 = \theta_1-g(\theta_1)/g'(\theta_1).$$

Then θ_2 is a better approximation to $\hat{\theta}$ than θ_1. We can improve on θ_2 in the same way and thus obtain an iterative numerical procedure defined by the sequence of approximations

$$\theta_{j+1} = \theta_j-g(\theta_j)/g'(\theta_j) \qquad (j = 2, 3, \ldots).$$

A variant of the procedure consists in replacing $g'(\theta)$ throughout the iterations by its expected value over all possible samples, namely $E(\partial^2 \log L/\partial\theta^2)$, which may be a simpler quantity to calculate.

Example 3. A sample of size n is taken from a distribution with density function

$$f(x;\theta) = \tfrac{1}{2}(1+\theta x) \qquad (-1 \leqslant x \leqslant 1, -1 \leqslant \theta \leqslant 1).$$

In example 3 of §**33**, we showed that

$$\partial \log l/\partial\theta = \Sigma\{x_i/(1+\theta x_i)\}.$$

Thus the maximum likelihood estimate of θ is a solution of the equation $g(\theta) = 0$, where

$$g(\theta) = \Sigma\{x_i/(1+\theta x_i)\}.$$

We take $\theta_1 = 3\,\Sigma x_i/n$ because $3M$ is an unbiased estimator of θ. In the iterative process, we can use either

$$-g'(\theta) = \Sigma\{x_i^2/(1+\theta x_i)^2\},$$

or
$$-E(\partial^2 \log L/\partial\theta^2) = n\left\{\log\left(\frac{1+\theta}{1-\theta}\right)-2\theta\right\}/2\theta^2.$$

The second function of θ is more easily calculated than the first, and therefore preferable.

Suppose next that a random sample is taken from a distribution which belongs to a family indexed by two parameters θ and β. The likelihood function is

$$l(\theta, \beta; x_1, x_2, \ldots, x_n) = f(x_1; \theta, \beta) f(x_2; \theta, \beta) \ldots f(x_n; \theta, \beta).$$

According to the principle of maximum likelihood, θ and β are estimated by $\hat{\theta}(x_1, x_2, \ldots, x_n)$ and $\hat{\beta}(x_1, x_2, \ldots, x_n)$ respectively, where the estimates are such that l is maximized when $\theta = \hat{\theta}$ and $\beta = \hat{\beta}$. If $\log l$ is differentiable, then the estimates can be obtained by solving the simultaneous equations

$$\partial \log l/\partial\theta = 0 \quad \text{and} \quad \partial \log l/\partial\beta = 0.$$

Example 4. A sample of size n is taken from a normal distribution with mean θ and variance β. We then have

$$\begin{aligned} l &= (2\pi\beta)^{-\frac{1}{2}n} \exp\{-\Sigma(x_i-\theta)^2/2\beta\}, \\ \log l &= -\tfrac{1}{2}n \log(2\pi) - \tfrac{1}{2}n \log \beta - \Sigma(x_i-\theta)^2/2\beta, \\ \partial \log l/\partial\theta &= \Sigma(x_i-\theta)/\beta, \end{aligned}$$

and

$$\partial \log l/\partial\beta = -n/2\beta + \Sigma(x_i-\theta)^2/2\beta^2.$$

The equations which $\hat{\theta}$ and $\hat{\beta}$ satisfy are

$$\Sigma(x_i-\hat{\theta})/\hat{\beta} = 0 \quad \text{and} \quad \Sigma(x_i-\hat{\theta})^2/2\hat{\beta}^2 = n/2\hat{\beta}.$$

From the first we obtain $\hat{\theta} = m$, and the second then gives

$$\hat{\beta} = \Sigma(x_i-m)^2/n = (n-1)s^2/n,$$

where m and s^2 are the sample mean and variance respectively. Since $(n-1)S^2/n$ is a biased estimator of σ^2, we conclude that the principle of maximum likelihood does not necessarily lead to unbiased estimators.

§36. Confidence intervals

We shall illustrate the procedures of statistical inference which are discussed in this section and the next by reference to a random sample from a normal distribution, known as a **normal sample**. The convenient shorthand $\mathcal{N}(\mu, \sigma^2)$ will signify a normal distribution with mean μ and variance σ^2, and describe any variate having this distribution. Thus a standard normal variate is $\mathcal{N}(0, 1)$. Suppose that a random sample of size n is taken from $\mathcal{N}(\mu, \sigma^2)$, and denote the corresponding variates by $X_1, X_2, \ldots, X_n$. The distribution of the sample mean $M = \Sigma X_i/n$ is obtained from example 3 of §**22** by taking

$$a_j = 1/n,\ \mu_j = \mu, \text{ and } \sigma_j^2 = \sigma^2 \qquad (j = 1, 2, \ldots, n).$$

We find that M is $\mathcal{N}(\mu, \sigma^2/n)$ and therefore $n^{\frac{1}{2}}(M-\mu)/\sigma$ is $\mathcal{N}(0, 1)$.

Consider an experiment where the observations can be regarded as a sample from a normal distribution with unknown mean but known variance. This situation arises, for example, if changes in the experimental conditions affect the population mean but not the population variance. From past records, the variance can then be estimated over long periods with unchanged conditions, and the resulting estimate identified with the true value. Formally, let the observations consist of a random sample of size n from $\mathcal{N}(\mu, \sigma^2)$, where μ is unknown but σ^2 is known. According to example 2 in §**33**, the sample mean M is a minimum variance bound estimator of μ when $\sigma^2 = 1$, and a similar argument shows that M has this property for arbitrary σ^2. However, more informative statements about μ can be made as follows.

Write u_ε for the upper 100ε percent point of a standard normal distribution. A table of u_ε is given below for the small values of ε that are usually specified.

100ε	0·05	0·5	2·5	5·0
u_ε	3·29	2·58	1·96	1·64

Suppose that ε_1 and ε_2 are probabilities such that $\varepsilon_1+\varepsilon_2 = \varepsilon$, where $\varepsilon < \frac{1}{2}$. Since $n^{\frac{1}{2}}(M-\mu)/\sigma$ is $\mathcal{N}(0, 1)$, we obtain

$$p\{-u_{\varepsilon_1} < n^{\frac{1}{2}}(M-\mu)/\sigma < u_{\varepsilon_2}\} = 1-\varepsilon,$$

where the condition on ε implies that both u_{ε_1} and u_{ε_2} are positive. The inequality $-u_{\varepsilon_1} < n^{\frac{1}{2}}(M-\mu)/\sigma$ can be expressed as $\mu < M+u_{\varepsilon_1}\sigma/n^{\frac{1}{2}}$, and similarly $n^{\frac{1}{2}}(M-\mu)/\sigma < u_{\varepsilon_2}$ implies that $\mu > M-u_{\varepsilon_2}\sigma/n^{\frac{1}{2}}$. Consequently

$$p(M-u_{\varepsilon_2}\sigma/n^{\frac{1}{2}} < \mu < M+u_{\varepsilon_1}\sigma/n^{\frac{1}{2}}) = 1-\varepsilon.$$

We have therefore converted a fixed interval for a variate $n^{\frac{1}{2}}(M-\mu)/\sigma$ into a random interval which covers the parameter μ with probability $1-\varepsilon$. The observed position of the random interval extends from $m-u_{\varepsilon_2}\sigma/n^{\frac{1}{2}}$ to $m+u_{\varepsilon_1}\sigma/n^{\frac{1}{2}}$, which are known as **confidence limits.** We say that the interval they enclose is a **confidence interval** for μ with **confidence coefficient** $1-\varepsilon$.

A small value of ε is chosen, so that $1-\varepsilon$ is close to 1. Thus the interpretation of a confidence interval is that in repeated samples the interval is almost certain to cover μ. For the sample we have obtained, either the interval covers μ or it fails to cover. Any reference to probabilities in connection with this particular sample would be inappropriate and another terminology has therefore been introduced. When ε_1 and ε_2 both differ from zero, the confidence limits are finite and the interval is known as **two-sided**. When $\varepsilon_1 = 0$, the interval is defined by $\mu > m-u_\varepsilon\sigma/n^{\frac{1}{2}}$, and is described as **one-sided**. Similarly, when $\varepsilon_2 = 0$, we obtain the one-sided interval $\mu < m+u_\varepsilon \sigma/n^{\frac{1}{2}}$.

The question arises as to how the values of ε_1 and ε_2 should be selected when $\varepsilon_1+\varepsilon_2 = \varepsilon$ is fixed. A natural solution is to minimize the length of the confidence interval, namely $(u_{\varepsilon_1}+u_{\varepsilon_2})\sigma/n^{\frac{1}{2}}$. The minimum occurs when $\varepsilon_1 = \varepsilon_2 = \frac{1}{2}\varepsilon$, a result which is readily conjectured from the shape of the normal distribution. We begin the proof by noting that one of the probabilities ε_1 and ε_2 must equal or exceed $\frac{1}{2}\varepsilon$. Suppose that $\varepsilon_1 = \frac{1}{2}\varepsilon+\delta$ where $\delta > 0$, in which case $\varepsilon_2 = \frac{1}{2}\varepsilon-\delta$. Then

$$0 < u_{\frac{1}{2}\varepsilon+\delta} < u_{\frac{1}{2}\varepsilon} < u_{\frac{1}{2}\varepsilon-\delta}.$$

The value of the density function $(2\pi)^{-\frac{1}{2}} \exp(-\frac{1}{2}x^2)$ throughout the interval $(u_{\frac{1}{2}\varepsilon+\delta}, u_{\frac{1}{2}\varepsilon})$ is everywhere larger than it is throughout the interval $(u_{\frac{1}{2}\varepsilon}, u_{\frac{1}{2}\varepsilon-\delta})$. Since the two intervals enclose the same probability δ, the first interval must be shorter than the second, so that

$$u_{\frac{1}{2}\varepsilon}-u_{\frac{1}{2}\varepsilon+\delta} < u_{\frac{1}{2}\varepsilon-\delta}-u_{\frac{1}{2}\varepsilon}.$$

An equivalent result is

$$(u_{\frac{1}{2}\varepsilon+\delta}+u_{\frac{1}{2}\varepsilon-\delta})\sigma/n^{\frac{1}{2}} > 2u_{\frac{1}{2}\varepsilon}\sigma/n^{\frac{1}{2}}.$$

Thus the length of the confidence interval is minimized when $\varepsilon_1 = \varepsilon_2 = \frac{1}{2}\varepsilon$.

Example. A civil engineer is erecting a power station which incorporates large quantities of concrete. He samples the mix from time to time and makes six-inch cubes which are tested for strength seven days later. The strength of a concrete cube is expressed in lb./sq. in., and records over a long period suggest that strength is normally distributed with standard deviation $\sigma = 400$. A further sample of 25 cubes using a different cement has a mean strength $m = 2950$, and σ is assumed to have the same value as before. The shortest confidence interval for μ with confidence coefficient 95 percent is $2950 \pm 1{\cdot}96(400)/25^{\frac{1}{2}}$. Thus the confidence limits are approximately 2790 and 3110. A one-sided confidence interval for μ with confidence coefficient 95 percent is defined by $\mu > 2950 - 1{\cdot}64(400)/25^{\frac{1}{2}}$, and thus consists of all values which exceed 2820, approximately.

§37. Tests of hypotheses

We have shown how the distribution of the sample mean provides a confidence interval for μ when the population variance is known. The following is another method of using the information contained in the distribution of the mean. Before the observations become available, we state that $\mu = \mu_0$, where μ_0 is a specified value which appears particularly relevant. For example, μ_0 is the mean life of an article which has been made by the same process for a long period, and the sample observations are the first of those on a cheaper process where the mean life may be different but the variance can be supposed the same. A statement which assigns specific values to unknown parameters of a probability distribution is described as a **statistical hypothesis**. Consider the hypothesis that $\mu = \mu_0$. In general, all values of μ such that $\mu \neq \mu_0$ are possible alternatives, but if, in the example described, those responsible for a new process are convinced that the mean life is at least as long as before, then the alternative to $\mu = \mu_0$ is $\mu > \mu_0$. The statement of a statistical hypothesis is accompanied by a statement of the **alternative** hypothesis because the best procedure to use may depend on the alternative, as will presently appear.

Denote the hypothesis that $\mu = \mu_0$ by $\mathscr{H}$ and let the alternative hypothesis be $\mu \neq \mu_0$. Taken together, they account for all the possible values of μ. We do not know whether $\mu = \mu_0$ or $\mu \neq \mu_0$. The type of argument adopted is to inspect the sample observations and then assert either $\mu = \mu_0$ or $\mu \neq \mu_0$, depending on the results obtained. An assertion that $\mu = \mu_0$ is described as **acceptance** of $\mathscr{H}$, and an assertion that $\mu \neq \mu_0$ as **rejection** of $\mathscr{H}$. Suppose that ε_1 and ε_2 are probabilities such that $\varepsilon_1 + \varepsilon_2 = \varepsilon$, where $\varepsilon < \frac{1}{2}$. Consider the procedure which consists of rejecting $\mathscr{H}$ whenever $n^{\frac{1}{2}}(m - \mu_0)/\sigma < -u_{\varepsilon_1}$ or $n^{\frac{1}{2}}(m - \mu_0)/\sigma > u_{\varepsilon_2}$. When $\mu = \mu_0$, $n^{\frac{1}{2}}(M - \mu_0)/\sigma$ is $\mathscr{N}(0, 1)$, so that

$$p(\mathscr{H} \text{ rejected} \mid \mu = \mu_0) = \varepsilon.$$

If $\mu = \mu_0$, we are wrong to reject $\mathscr{H}$, and we wish to minimize

$p(\mathscr{H}$ rejected $| \mu = \mu_0)$. A small value of ε is therefore chosen, usually 0·05 or 0·01.

The choice of ε_1 and ε_2 affects what happens when $\mu \neq \mu_0$. When the population mean is μ, M is $\mathscr{N}(\mu, \sigma^2/n)$ and $n^{\frac{1}{2}}(M-\mu_0)/\sigma$ is $\mathscr{N}\{n^{\frac{1}{2}}(\mu-\mu_0)/\sigma, 1\}$. Consequently

$$p(\mathscr{H} \text{ rejected} \mid \mu) = p\{n^{\frac{1}{2}}(M-\mu_0)/\sigma < u_{\varepsilon_1} \mid \mu\} + p\{n^{\frac{1}{2}}(M-\mu_0)/\sigma > u_{\varepsilon_2} \mid \mu\}.$$

The second term on the right is

$$1-\Phi\{u_{\varepsilon_2}-n^{\frac{1}{2}}(\mu-\mu_0)/\sigma\} = \Phi\{-u_{\varepsilon_2}+n^{\frac{1}{2}}(\mu-\mu_0)/\sigma\}$$

and hence

$$p(\mathscr{H} \text{ rejected} \mid \mu) = \Phi\{-u_{\varepsilon_1}-n^{\frac{1}{2}}(\mu-\mu_0)/\sigma\} + \Phi\{-u_{\varepsilon_2}+n^{\frac{1}{2}}(\mu-\mu_0)/\sigma\}.$$

This takes the value ε when $\mu = \mu_0$. We select ε_1 and ε_2 such that the probability of rejection is a minimum when $\mu = \mu_0$. The derivative of $p(\mathscr{H}$ rejected $| \mu)$ with respect to μ is

$$(n^{\frac{1}{2}}/\sigma)[-\phi\{-u_{\varepsilon_1}-n^{\frac{1}{2}}(\mu-\mu_0)/\sigma\} + \phi\{-u_{\varepsilon_2}+n^{\frac{1}{2}}(\mu-\mu_0)/\sigma\}],$$

which is zero when $\mu = \mu_0$ if $\phi(-u_{\varepsilon_1}) = \phi(-u_{\varepsilon_2})$, or equivalently if $\varepsilon_1 = \varepsilon_2$. Since the derivative of $\phi(x)$ is $-x\phi(x)$, the second deriva-

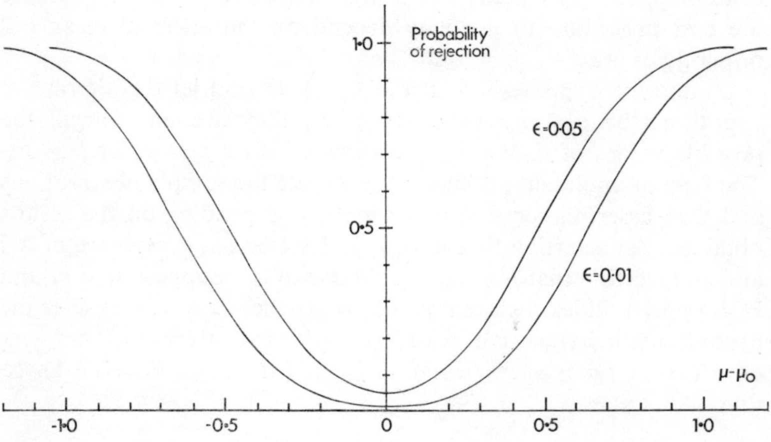

Fig. 11. Probability that test procedure $\mathscr{T}$ rejects $\mu = \mu_0$ as a function of $\mu - \mu_0$ when $\sigma = 1$, $n = 16$, $\epsilon = 0{\cdot}05$ and $\epsilon = 0{\cdot}01$.

tive of $p(\mathscr{H}$ rejected $| \mu)$ at $\mu = \mu_0$ is positive, which confirms that a minimum is attained. We therefore continue with a procedure for which $\varepsilon_1 = \varepsilon_2 = \frac{1}{2}\varepsilon$. The probability of rejection now increases with $|\mu - \mu_0|$ and tends to 1 as $|\mu - \mu_0| \to \infty$.

If $\mu \neq \mu_0$, we are right to reject $\mathscr{H}$, and we wish to maximize $p(\mathscr{H}$ rejected $| \mu \neq \mu_0)$. Unfortunately, this requirement runs counter to that of minimizing $p(\mathscr{H}$ rejected $| \mu = \mu_0)$, as Fig. 11 illustrates. The probability that $\mathscr{H}$ is rejected is displayed here as a function of $(\mu - \mu_0)$ for $\sigma = 1$, $n = 16$, and $\varepsilon = 0{\cdot}05, 0{\cdot}01$. Since the graph for $\varepsilon = 0{\cdot}01$ is everywhere below the graph for $\varepsilon = 0{\cdot}05$, a smaller probability of rejecting $\mathscr{H}$ when $\mu = \mu_0$ implies a smaller probability of rejecting $\mathscr{H}$ for any specified $\mu \neq \mu_0$. When choosing ε, we therefore need to compromise between the conflicting requirements of minimizing the first probability and maximizing the second. One method is to assign costs to wrong decisions, but we have no space for details.

The procedure of rejecting $\mathscr{H}$ when $|n^{\frac{1}{2}}(m - \mu_0)/\sigma| > u_{\frac{1}{2}\varepsilon}$ is known as a **test** of $\mathscr{H}$ at **significance level** ε, and $n^{\frac{1}{2}}(m - \mu_0)/\sigma$ is the **test statistic**. We can summarize the properties of the procedure for all values of μ by a graph of $p(\mathscr{H}$ rejected $| \mu)$, which is known as the **power** function of the test. Because the power function has a minimum when $\mu = \mu_0$, we say that the test is **unbiased**.

The test procedure has been formalized as a rigid set of rules in order that its properties can be determined precisely. In practice, a relaxed version is customary. A probability δ is calculated from

$$|n^{\frac{1}{2}}(m - \mu_0)/\sigma| = u_{\frac{1}{2}\delta},$$

and the data are said to be **significant** at the 100δ percent level. The value of δ is then assessed according to a scale where 0·05 is regarded as just indicative of a real difference $(\mu - \mu_0)$, and 0·01 is regarded as good evidence of a change in μ. Such a scale is inevitably subjective.

Example 1. The strength of six-inch concrete cubes is measured in lb./sq. in. seven days after the cubes are made. Past records indicate a normal distribution with $\mu = 2770$ and $\sigma = 400$. The mean of a sample of 25 cubes using a different cement is 2950, and σ is assumed to have the same value as before. We require to test the hypothesis that the population mean also is unchanged i.e. $\mu = 2770$. The value of the test statistic is

$$25^{\frac{1}{2}}(2950 - 2770)/400 = 2{\cdot}25,$$

giving $\frac{1}{2}\delta = 0{\cdot}0122$ and corresponding to a significance level of 2·44 percent. There is evidence of a real change in the mean strength of the concrete.

We now discuss how the hypothesis $\mathscr{H}: \mu = \mu_0$ can be tested when the alternative hypothesis is $\mu > \mu_0$ and the possibility that $\mu < \mu_0$ is excluded. Consider the test which rejects $\mathscr{H}$ whenever $n^{\frac{1}{2}}(m-\mu_0)/\sigma > u_\varepsilon$. Denote the previous test by $\mathscr{T}$ and the proposed test by $\mathscr{T}_+$. As before, the distribution of $n^{\frac{1}{2}}(M-\mu_0)/\sigma$ is $\mathscr{N}\{n^{\frac{1}{2}}(\mu-\mu_0)/\sigma, 1\}$, so that the power function of $\mathscr{T}_+$ is

$$p(\mathscr{H} \text{ rejected} \mid \mu) = \Phi\{-u_\varepsilon + n^{\frac{1}{2}}(\mu-\mu_0)/\sigma\}.$$

The probability of rejection is ε when $\mu = \mu_0$, decreases to 0 as $\mu \to -\infty$, and increases to 1 as $\mu \to \infty$. Fig. 12 displays the power

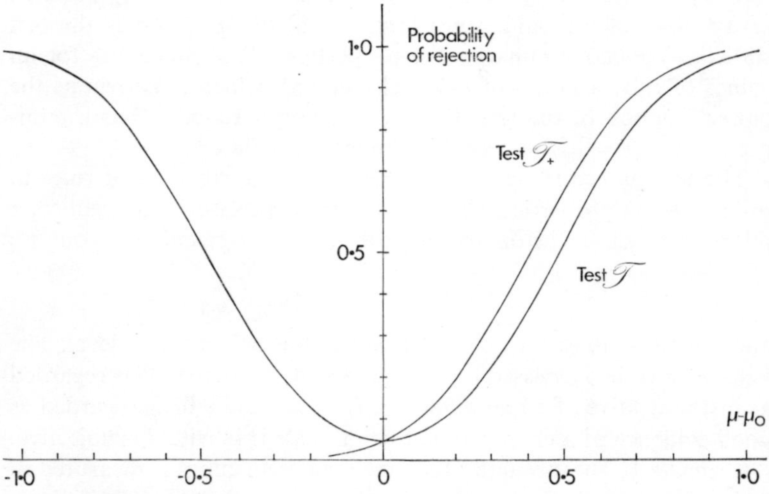

Fig. 12. Probability that test procedures $\mathscr{T}$ and $\mathscr{T}_+$ reject $\mu = \mu_0$ as a function of $\mu - \mu_0$ when $\sigma = 1$, $n = 16$, $\epsilon = 0{\cdot}05$.

function of $\mathscr{T}_+$ when $\sigma = 1$, $n = 16$, and $\varepsilon = 0{\cdot}05$, together with the power function of $\mathscr{T}$ for the same values of σ, n, ε. The test procedure $\mathscr{T}_+$ is more powerful than $\mathscr{T}$ for $\mu > \mu_0$ in the sense that the power function of $\mathscr{T}_+$ lies above that of $\mathscr{T}$ for all μ in this range.

On the other hand, $\mathscr{T}_+$ is less powerful for $\mu < \mu_0$, and indeed the probability of rejecting $\mathscr{H}$ is less than for $\mu = \mu_0$. This region of μ is strictly not relevant for $\mathscr{T}_+$ but the graph clearly shows the consequences of mistakenly excluding the possibility that $\mu < \mu_0$. We say that $\mu > \mu_0$ is a **one-sided** alternative, whereas $\mu \neq \mu_0$ is a **two-sided** alternative.

Example 2. For the data of Example 1, $m = 2950$, $\sigma = 400$, and $n = 25$. The accepted value of μ is 2770, and when the different cement is used, a mean strength below this level is not considered possible. Here $u_\delta = 2{\cdot}25$, and the data are significant at a level of 1·22 percent, half that of Example 1. There is now strong evidence of a change in the mean strength.

A simple connection holds between significance tests and confidence intervals. Consider the unbiased test of $\mathscr{H}: \mu = \mu_0$ when the alternative hypothesis is $\mu \neq \mu_0$. $\mathscr{H}$ is rejected if $|n^{\frac{1}{2}}(m-\mu_0)/\sigma| > y_{\frac{1}{2}\varepsilon}$, in which case either $\mu_0 > m+u_{\frac{1}{2}\varepsilon}\sigma/n^{\frac{1}{2}}$ or $\mu_0 < m-u_{\frac{1}{2}\varepsilon}\sigma/n^{\frac{1}{2}}$, so that the confidence interval $m \pm u_{\frac{1}{2}\varepsilon}\sigma/n^{\frac{1}{2}}$ fails to contain μ_0. The test which rejects $\mathscr{H}$ when $n^{\frac{1}{2}}(m-\mu_0)/\sigma > u_\varepsilon$ and the confidence interval $(m-u_\varepsilon\sigma/n^{\frac{1}{2}}, \infty)$ are similarly related. In general, therefore, the hypothesis $\mu = \mu_0$ is rejected at significance level ε whenever μ_0 lies outside a confidence interval for μ with confidence coefficient $1-\varepsilon$.

EXERCISES 7

1. The probability is θ that an inspected item is defective. Items are inspected one by one until n defectives are found, and the total number of items inspected is then Y. Prove that

$$E\{(n-1)/(Y-1)\} = \theta \quad \text{and} \quad E\{(n-1)(n-2)/(Y-1)(Y-2)\} = \theta^2.$$

Write $R = (n-1)/(Y-1)$. Deduce that R is an unbiased estimator of θ, and that an unbiased estimator of var (R) is $R(1-R)/(Y-2)$.

2. A sample of size n is taken from a Poisson distribution with parameter θ. The sample mean is M. Write $y = n \log \{n/(n-1)\}$, and $U = e^{-yM}$. Show that U is the unique unbiased estimator of $e^{-\theta}$ with minimum variance, and that

$$\text{var}\,(U) = \exp\{-\theta(2n-1)/n\} - \exp(-2\theta).$$

Another unbiased estimator of $e^{-\theta}$ is $T = N_0/n$, where N_0 is the number of zeros in the sample. Prove that

$$\text{var}\,(U)/\text{var}\,(T) = n/\{1+e^{\theta/n}+e^{2\theta/n}+\ldots+e^{(n-1)\theta/n}\}.$$

3. A truncated Poisson distribution is defined by

$$p(X = x) = \theta^x/x\,!\,(e^\theta - 1) \quad (x = 1, 2, 3, \ldots).$$

In a sample of size n from this distribution, there are N_v observations which have the value v ($v = 1, 2, 3, \ldots$). Write

$$T = \sum_{v=2}^{\infty} v\,N_v/n.$$

Prove that T is an unbiased estimator of θ and that

$$\text{var}\,(T) = \{\theta + \theta^2/(e^\theta - 1)\}/n.$$

Show that a lower bound for the variance of an unbiased estimator is given by

$$\theta(1 - e^{-\theta})^2/n(1 - e^{-\theta} - \theta e^{-\theta}).$$

Calculate the relative efficiency of T and show that it never falls below the value 0·9536, attained when $\theta = 1{\cdot}355$.

4. A sample of size n is taken from the binomial distribution

$$f(x; \theta) = \binom{k}{x} \theta^x (1-\theta)^{k-x} \qquad (x = 0, 1, 2, \ldots, k).$$

The sample mean is M. Show that M/k is a minimum variance bound estimator of θ with variance $\theta(1-\theta)/nk$.

5. A sample of size n is taken from an exponential distribution with parameter λ. The sample mean is M. Show that the maximum likelihood estimator of λ is M^{-1}.

The distribution of nM is given in exercise 5.3. Deduce that $(n-1)/nM$ is an unbiased estimator of λ with variance $\lambda^2/(n-2)$.

6. The **double exponential** or **Laplace** distribution is defined by the density function

$$f(x; \theta, \beta) = (2\beta)^{-1} \exp\,(-|\,x - \theta\,|/\beta),$$

where $-\infty < x < \infty$, $-\infty < \theta < \infty$, and $\beta > 0$. Show that the maximum likelihood estimates of θ and β from a sample of size n are

$$\hat{\theta} = q \quad \text{and} \quad \hat{\beta} = \Sigma\,|\,x_i - q\,|/n,$$

where q is the sample median.

7. A sample of size n is taken from the distribution with density function

$$f(x) = \begin{cases} \theta^{-1} & (0 < x < \theta), \\ 0 & \text{otherwise}. \end{cases}$$

Denote the maximum sample value by x_{nn}. Show that the joint density function of $X_1, X_2, \ldots, X_n$ is θ^{-n} when $x_{nn} < \theta$ and zero when $x_{nn} \geqslant \theta$. Sketch the likelihood function and deduce that the maximum likelihood estimate of θ is x_{nn}. Notice that assumptions (*ii*) and (*iii*) of §**32** are not satisfied for this distribution.

8. The conditions are the same as in the previous exercise. Prove that $(x_{nn}, x_{nn}\,\varepsilon^{-1/n})$ is a confidence interval for θ with confidence coefficient $1-\varepsilon$.

Let W be the sample range. The distribution of W/θ is given in exercise 6.9. Denote by $\Omega_{1-\varepsilon}$ the lower 100ε percent point of this distribution. Show that $(w, w/\Omega_{1-\varepsilon})$ is a confidence interval for θ with confidence coefficient $1-\varepsilon$.

When $n = 5$ and $\varepsilon = 0{\cdot}05$, compare the expected lengths of the two intervals.

9. A sample of size n is taken from the distribution of a continuous variate X. The order statistics are denoted by $x_{1n}, x_{2n}, \ldots, x_{nn}$. Let ξ_ε be the upper 100ε percent point of the distribution of X. For $j < k$, show that (x_{jn}, x_{kn}) is a confidence interval for ξ_ε with confidence coefficient

$$\sum_{i=j}^{k-1} \binom{n}{i} (1-\varepsilon)^i\, \varepsilon^{n-i}.$$

Hence obtain a test of the hypothesis $\xi_\varepsilon = c$ when the alternative hypothesis is $\xi_\varepsilon \neq c$.

10. A sample of size n is taken from a normal distribution with mean zero and variance θ. Show that $\Sigma X_i^2/n$ is a minimum variance bound estimator of θ and is distributed as $\theta\chi_n^2/n$.

Denote the distribution function of χ_n^2 by Q_n and let $\chi^2_{n;1-\varepsilon}$ be the lower 100ε percent point of χ_n^2. A test of the hypothesis $\mathscr{H}: \theta = \theta_0$ against the alternative $\theta > \theta_0$ is defined by rejecting $\mathscr{H}$ whenever $\Sigma X_i^2 < \theta_0\chi^2_{n;1-\varepsilon}$. Show that the significance level is ε and that the power function is given by $Q_n(\theta_0\chi^2_{n;1-\varepsilon}/\theta)$.

Obtain a test of $\mathscr{H}$ against the alternative $\theta \neq \theta_0$ and express its power function in terms of the distribution function and percentage points of χ_n^2.

CHAPTER 8
NORMAL SAMPLES

§38. One normal sample

The statistical model in the previous two sections is defined by a random sample from a normal distribution with unknown mean but known variance σ^2. This model provides a convenient introduction to confidence intervals and tests of hypotheses, but its practical applications are severely limited by the assumption that σ^2 is known. Various generalizations of the model are useful in practice, and this chapter is concerned with problems of statistical inference which then arise.

Consider first a model defined by a random sample from a normal distribution when both the mean and variance are unknown. We have shown that the sample mean is normally distributed, but an analysis of the data now depends on two further properties of a normal sample. First, the sample mean and variance are independent, and secondly, the sample variance is distributed as a multiple of χ^2. The proofs are as follows.

Let $X_1, X_2, \ldots, X_n$ be the variates observed in a sample of size n from $\mathcal{N}(\mu, \sigma^2)$. Denote the sample mean and variance by M_x and S_x^2 respectively. Thus

$$M_x = \Sigma X_j/n \quad \text{and} \quad S_x^2 = \Sigma(X_j - M_x)^2/(n-1).$$

Define

$$Y_j = (X_j - \mu)/\sigma \qquad (j = 1, 2, \ldots, n).$$

Then $Y_1, Y_2, \ldots, Y_n$ are the variates corresponding to a sample of size n from $\mathcal{N}(0, 1)$. Denote the sample mean and variance by M_y and S_y^2 respectively. Thus

$$M_y = \Sigma Y_j/n \quad \text{and} \quad S_y^2 = \Sigma(Y_j - M_y)^2/(n-1).$$

The relationships

$$M_x = \mu + \sigma M_y \quad \text{and} \quad S_x^2 = \sigma^2 S_y^2$$

are derived in the same way as the computing formulae of §**29**. Consider an orthonormal transformation from $Y_1, Y_2, \ldots, Y_n$ to $T_1, T_2, \ldots, T_n$ in which

$$T_1 = n^{-\frac{1}{2}}(Y_1 + Y_2 + \ldots + Y_n).$$

Equivalently $T_1 = n^{\frac{1}{2}} M_y$, and hence

$$(n-1)S_y^2 = \sum_{j=1}^{n} Y_j^2 - nM_y^2 = \sum_{i=1}^{n} T_i^2 - T_1^2.$$

This gives the representation

$$(n-1)S_y^2 = \sum_{i=2}^{n} T_i^2.$$

According to example 3 of §**22**, $T_1, T_2, \ldots, T_n$ are mutually independent $\mathcal{N}(0, 1)$ variates. Therefore M_y, which is a function of T_1 only, is independent of S_y^2. Since M_x is a function of M_y, and S_x^2 is a function of S_y^2, we infer that M_x and S_x^2 are also independent. Thus the mean and variance of a sample from $\mathcal{N}(\mu, \sigma^2)$ are independent.

The distinction between S_x^2 and S_y^2 has now served its purpose. Let S^2 be the variance of a sample of size n from $\mathcal{N}(\mu, \sigma^2)$. Then

$$S^2 = \frac{\sigma^2}{n-1} \sum_{i=2}^{n} T_i^2.$$

We have therefore shown that the sample variance is distributed as $\chi_{n-1}^2 \sigma^2/(n-1)$.

Let A and B be independent variates where A is $\mathcal{N}(0, 1)$ and B is χ_ν^2. Define

$$T = A/(B/\nu)^{\frac{1}{2}}.$$

The density function of T can be found by the methods of §**22** (see exercise 4). We say that T has 'Student's' t-distribution with ν degrees of freedom. The distribution is named after its discoverer W. S. Gosset, who published under the pseudonym of 'Student'. It is symmetrical about zero, and similar in appearance to a standard normal distribution, but the density function tends to zero less rapidly as $t \to \pm\infty$. Denote the upper 100ε percent point by $t_{\nu;\,\varepsilon}$ so that $t_{\nu;\,1-\varepsilon} = -t_{\nu;\,\varepsilon}$. A table of percentage points is given by Pearson and Hartley (1966).

Suppose that a sample of size n is taken from $\mathcal{N}(\mu, \sigma^2)$ when both μ and σ^2 are unknown. We want to derive confidence intervals and significance tests for μ. Inferences about μ when σ^2 is known are based on the fact that $n^{\frac{1}{2}}(M-\mu)/\sigma$ is $\mathcal{N}(0, 1)$. When σ^2 is unknown, a natural modification is to replace σ by the sample standard deviation and consider the variate $n^{\frac{1}{2}}(M-\mu)/S$. This can be expressed as the ratio of $n^{\frac{1}{2}}(M-\mu)/\sigma$ to $(S^2/\sigma^2)^{\frac{1}{2}}$, where the denominator is independent of the numerator and such that $(n-1)S^2/\sigma^2$ is distributed as χ^2_{n-1}.

By taking $A = n^{\frac{1}{2}}(M-\mu)/\sigma$, $B = (n-1)S^2/\sigma^2$, and $\nu = n-1$, we find that $n^{\frac{1}{2}}(M-\mu)/S$ has 'Student's' t-distribution with $(n-1)$ degrees of freedom. Hence

$$p\{-t_{n-1;\,\frac{1}{2}\varepsilon} < n^{\frac{1}{2}}(M-\mu)/S < t_{n-1;\,\frac{1}{2}\varepsilon}\} = 1-\varepsilon.$$

When the inequalities on the left are rearranged,

$$p(M-t_{n-1;\,\frac{1}{2}\varepsilon}S/n^{\frac{1}{2}} < \mu < M+t_{n-1;\,\frac{1}{2}\varepsilon}S/n^{\frac{1}{2}}) = 1-\varepsilon.$$

Thus $m \pm t_{n-1;\,\frac{1}{2}\varepsilon}s/n^{\frac{1}{2}}$ is a confidence interval for μ with confidence coefficient $1-\varepsilon$. The hypothesis $\mathscr{H}: \mu = \mu_0$ can be tested at significance level ε by the procedure which consists in rejecting $\mathscr{H}$ whenever μ_0 lies outside the confidence interval. Similar remarks apply in respect of the one-sided confidence intervals

$$(-\infty, m+t_{n-1;\,\varepsilon}s/n^{\frac{1}{2}}) \quad \text{and} \quad (m-t_{n-1;\,\varepsilon}s/n^{\frac{1}{2}}, \infty).$$

Example. The results given below come from an experiment to find whether it was of advantage to kiln-dry barley seed before sowing ('Student', 1908). Eleven varieties of barley were sown, and for each variety the difference is recorded between the kiln-dried and not kiln-dried in lb. head corn per acre.

$$+106,\ -20,\ +101,\ -33,\ +72,\ -36,\ +62,\ +38,\ -70,\ +127,\ +24.$$

We assume that the observations form a sample of 11 from $\mathcal{N}(\mu, \sigma^2)$ where σ^2 is unknown. Our object is to test the hypothesis $\mu = 0$ when the alternative is $\mu > 0$. The test statistic is $t = n^{\frac{1}{2}}m/s$, and we reject $\mu = 0$ if $t > t_{10;\,\varepsilon}$ for ε sufficiently small. Here $m = 33{\cdot}7$, $s = 66{\cdot}17$, and $t = 1{\cdot}52$. Since $t_{10;\,0{\cdot}05} = 1{\cdot}81$ and $t_{10;\,0{\cdot}10} = 1{\cdot}37$, the data are significant at a level between 5 and 10 percent. There is no particular advantage in kiln-drying before sowing.

§39. Two normal samples

The notion of independent samples of size n from a single distribution was introduced in §**27**. A straightforward generalization is that samples of sizes $n_1, n_2, \ldots$ from different distributions are independent if the $n_1+n_2+\ldots$ variates observed are mutually independent. The corresponding sample means are then mutually independent and similarly for any other statistic. In this section, we suppose that the data consist of independent samples of sizes n_1 and n_2 taken from $\mathcal{N}(\mu_1, \sigma^2)$ and $\mathcal{N}(\mu_2, \sigma^2)$ respectively, where μ_1, μ_2 and σ^2 are unknown. We want to derive confidence intervals and significance tests for $\mu_1-\mu_2$.

Denote the sample means by M_1 and M_2 respectively. Define

$$A = \{(M_1-M_2)-(\mu_1-\mu_2)\}\{\sigma^2(n_1^{-1}+n_2^{-1})\}^{-\frac{1}{2}}.$$

By repeated applications of the final result in example 3 of §**22**, we find that M_1 is $\mathcal{N}(\mu_1, \sigma^2/n_1)$, M_2 is $\mathcal{N}(\mu_2, \sigma^2/n_2)$ and M_1-M_2 is $\mathcal{N}(\mu_1-\mu_2, \sigma^2/n_1+\sigma^2/n_2)$. Consequently A is $\mathcal{N}(0, 1)$.

Denote the sample variances by S_1^2 and S_2^2 respectively. Define

$$B = \{(n_1-1)S_1^2+(n_2-1)S_2^2\}/\sigma^2.$$

Since $(n_1-1)S_1^2/\sigma^2$ and $(n_2-1)S_2^2/\sigma^2$ are independent and distributed as $\chi^2_{n_1-1}$ and $\chi^2_{n_2-1}$ respectively, their total is $\chi^2_{n_1+n_2-2}$. Furthermore, the variates M_1, M_2, S_1^2, and S_2^2 are mutually independent, and hence A is distributed independently of B. Define

$$T = \frac{A}{\left(\dfrac{B}{n_1+n_2-2}\right)^{\frac{1}{2}}} = \frac{(M_1-M_2)-(\mu_1-\mu_2)}{\left\{\dfrac{(n_1-1)S_1^2+(n_2-1)S_2^2}{n_1+n_2-2}\left(\dfrac{1}{n_1}+\dfrac{1}{n_2}\right)\right\}^{\frac{1}{2}}}.$$

We conclude that T has 'Student's' t-distribution with n_1+n_2-2 degrees of freedom. The expression for T in terms of the sample means and variances has the same structure as $n^{\frac{1}{2}}(M-\mu)/S$, and an argument similar to the one used to derive confidence limits for μ shows that

$$(m_1-m_2) \pm t_{n_1+n_2-2;\,\frac{1}{2}\varepsilon}\left\{\frac{(n_1-1)s_1^2+(n_2-1)s_2^2}{n_1+n_2-2}\left(\frac{1}{n_1}+\frac{1}{n_2}\right)\right\}^{\frac{1}{2}}$$

is a confidence interval for $\mu_1-\mu_2$ with confidence coefficient $1-\varepsilon$.

Equivalently, the hypothesis $\mathscr{H}: \mu_1 - \mu_2 = 0$ can be tested at significance level ε by rejecting $\mathscr{H}$ whenever the confidence interval fails to cover zero.

Example. Experiments were made by the third Lord Rayleigh (Strutt, 1903) to determine the density of nitrogen prepared in various ways, and the following data give the weights of nitrogen in a standard bulb. In the first series of determinations, the nitrogen was prepared by passing air over hot copper; in the second series, air was passed over hot iron.

Reducing agent			*Weight of nitrogen*		
Copper	2·31035	2·31026	2·31024	2·31012	2·31027
Iron	2·31017	2·30986	2·31010	2·31001	

We assume that the first series represents a sample of 5 from $\mathscr{N}(\mu_1, \sigma^2)$ and the second series a sample of 4 from $\mathscr{N}(\mu_2, \sigma^2)$ where the common variance σ^2 is unknown. Our object is to test $\mu_1 = \mu_2$ when the alternative is $\mu_1 \neq \mu_2$. The value of the test statistic is unchanged by a linear transformation applied to μ_1, μ_2 and the observations. We therefore work in units of 10^{-5} referred to an origin of 2·31, and find

$$m_1 = 24{\cdot}8,\ m_2 = 3{\cdot}5,\ (n_1-1)s_1^2 = 275,\ (n_2-1)s_2^2 = 537,$$

$$t = \frac{24{\cdot}8 - 3{\cdot}5}{\left\{\dfrac{812}{7}\left(\dfrac{1}{4}+\dfrac{1}{5}\right)\right\}^{\frac{1}{2}}} = 2{\cdot}95.$$

Here $\nu = 7$ and the data are significant at a level close to 2 percent. Thus the two methods of preparation lead to gases of a slightly different composition, and the reasons must be sought in a careful analysis of the experimental details.

§40. Several normal samples

We now suppose that the data consist of independent samples of sizes $n_1, n_2, \ldots, n_k$ taken from $\mathscr{N}(\mu_1, \sigma^2), \mathscr{N}(\mu_2, \sigma^2), \ldots, \mathscr{N}(\mu_k, \sigma^2)$ respectively, where $\mu_1, \mu_2, \ldots, \mu_k$ and σ^2 are unknown. The problem is to test the hypothesis

$$\mathscr{H}: \mu_1 = \mu_2 = \ldots = \mu_k.$$

Denote the jth member of the ith sample by x_{ij} and the corresponding variate by X_{ij}, where $j = 1, 2, \ldots, n_i$ and $i = 1, 2, \ldots, k$.

We require the likelihood of the values $\{x_{ij}\}$, say l. The definition of likelihood given in §**32** is applied there when the variates observed are independently and identically distributed. However, the definition is valid whatever the joint distribution of the variates, and so l is the joint density function of $\{X_{ij}\}$ evaluated at $\{x_{ij}\}$ respectively. Thus

$$l = \prod_{i=1}^{k} l_i \text{ where } l_i = (2\pi\sigma^2)^{-\frac{1}{2}n_i} \exp\left\{-\sum_{i=1}^{n_i} (x_{ij}-\mu_i)^2/2\sigma^2\right\}.$$

We therefore obtain

$$\log l = -\sum_{i=1}^{k} \tfrac{1}{2}n_i \log(2\pi\sigma^2) - \sum_{i=1}^{k}\sum_{j=1}^{n_i} (x_{ij}-\mu_i)^2/2\sigma^2.$$

The maximum likelihood estimates of $\mu_1, \mu_2, \ldots, \mu_k$ and σ^2 satisfy

$$\partial \log l/\partial\mu_i = 0 \quad (i = 1, 2, \ldots, k) \quad \text{and} \quad \partial \log l/\partial\sigma^2 = 0.$$

Since μ_i does not occur in l_h where $h \neq i$,

$$\partial \log l/\partial\mu_i = \partial \log l_i/\partial\mu_i.$$

Proceeding as in example 4 of §**35**, we find that the maximum likelihood estimate of μ_i is the ith sample mean, say m_i. Write n for Σn_i and consider next

$$\partial \log l/\partial\sigma^2 = -n/2\sigma^2 + \sum_{i=1}^{k}\sum_{j=1}^{n_i} (x_{ij}-\mu_i)^2/2(\sigma^2)^2.$$

Substitute the estimates of $\mu_1, \mu_2, \ldots, \mu_k$ in this derivative and solve the equation $\partial \log l/\partial\sigma^2 = 0$. The maximum likelihood estimate of σ^2 is z^2, where

$$z^2 = \sum_{i=1}^{k}\sum_{j=1}^{n_i} (x_{ij}-m_i)^2/n.$$

When $\mathscr{H}$ is true, let ξ be the common value of $\mu_1, \mu_2, \ldots, \mu_k$. We now have a sample of size n from $\mathcal{N}(\xi, \sigma^2)$. The maximum likelihood estimates of ξ and σ^2 are respectively m and z_0^2, where

$$m = \sum_{i=1}^{k}\sum_{j=1}^{n_i} x_{ij}/n = \sum_{i=1}^{k} n_i m_i/n,$$

and

$$z_0^2 = \sum_{i=1}^{k}\sum_{j=1}^{n_i} (x_{ij}-m)^2/n.$$

We see that the estimate of σ^2 depends on whether or not $\mathscr{H}$ is true.

In what follows, the notation is

$$w = \sum_{i=1}^{k} \sum_{j=1}^{n_i} (x_{ij}-\mu_i)^2, \quad r = \sum_{i=1}^{k} \sum_{j=1}^{n_i} (x_{ij}-m_i)^2,$$

and
$$r_0 = \sum_{i=1}^{k} \sum_{j=1}^{n_i} (x_{ij}-m)^2.$$

The procedure of maximizing log l with respect to $\mu_1, \mu_2, \ldots, \mu_k$ is equivalent to minimizing w with respect to the same parameters. We have shown that the overall minimum of w is r, and that when $\mathscr{H}$ is true the minimum of w is r_0. The second minimum is obtained by restricting the variation of $\mu_1, \mu_2, \ldots, \mu_k$ and therefore $r_0 \geqslant r$. Consider the difference between the two minima, namely

$$q = r_0 - r.$$

The difference q is known as the **sum of squares** for testing the hypothesis $\mathscr{H}$. We now derive properties of q which justify its use in this connection.

Let the variates corresponding to $m_i, q, r, \ldots$ be $M_i, Q, R, \ldots$ respectively. We proceed to find $E(R)$ and $E(Q)$ by the methods of §**28**. Write

$$R = \sum_{i=1}^{k} R_i \quad \text{where} \quad R_i = \sum_{j=1}^{n_i} (X_{ij}-M_i)^2.$$

Then $R_i/(n_i-1)$ is the ith sample variance, whence

$$E(R) = \sum_{i=1}^{k} (n_i-1)\sigma^2 = (n-k)\sigma^2.$$

Denote $E(M)$ by μ, so that $\mu = \Sigma n_i\mu_i/n$. Since

$$R_0 = \sum_{i=1}^{k} \sum_{j=1}^{n_i} X_{ij}^2 - nM^2$$

and
$$R = \sum_{i=1}^{k} \sum_{j=1}^{n_i} X_{ij}^2 - \sum_{i=1}^{k} n_i M_i^2,$$

another expression for Q is

$$Q = \sum_{i=1}^{k} n_i M_i^2 - nM^2.$$

From this version, we obtain

$$E(Q) = \sum_{i=1}^{k} n_i(\mu_i^2+\sigma^2/n_i)-n(\mu^2+\sigma^2/n)$$

$$= (k-1)\sigma^2+\sum_{i=1}^{k} n_i(\mu_i-\mu)^2.$$

These results permit the following conclusions. When $\mathscr{H}$ is true, $Q/(k-1)$ is an unbiased estimator of σ^2, but otherwise $Q/(k-1)$ tends to overestimate σ^2, because $\Sigma n_i(\mu_i-\mu)^2 \geqslant 0$ with equality if and only if $\mu_i = \xi$ for $i = 1, 2, \ldots, k$. On the other hand, $R/(n-k)$ is an unbiased estimator of σ^2 whether $\mathscr{H}$ is true or not. Thus the ratio of $Q/(k-1)$ to $R/(n-k)$ is an appropriate statistic for testing $\mathscr{H}$. Values around 1 occur when $\mathscr{H}$ is true and large values suggest that $\mathscr{H}$ is false. We say that $Q/(k-1)$ and $R/(n-k)$ are **mean squares**, while $\{Q/(k-1)\}/\{R/(n-k)\}$ is called a **variance-ratio**.

The final stage in constructing a test of $\mathscr{H}$ is to determine the distribution of the variance-ratio when $\mathscr{H}$ is true. We begin by showing that Q/σ^2 is χ^2_{k-1}. The version

$$Q = \sum_{i=1}^{k} n_i(M_i-\xi)^2-n(M-\xi)^2$$

is identically equal to the forms of Q already given. Write

$$Y_i = n_i^{\frac{1}{2}}(M_i-\xi)/\sigma \qquad (i = 1, 2, \ldots, k),$$

so that Y_i is $\mathscr{N}(0, 1)$. Make an orthonormal transformation from $Y_1, Y_2, \ldots, Y_k$ to $T_1, T_2, \ldots, T_k$, such that

$$T_1 = \sum_{i=1}^{k} (n_i/n)^{\frac{1}{2}} Y_i.$$

Equivalently, $T_1 = \Sigma n_i(M_i-\xi)/n^{\frac{1}{2}}\sigma = n^{\frac{1}{2}}(M-\xi)/\sigma$, and hence

$$Q = \sigma^2\left(\sum_{i=1}^{k} Y_i^2-T_1^2\right) = \sigma^2 \sum_{j=2}^{k} T_j^2.$$

This shows that Q/σ^2 is χ^2_{k-1}. We turn next to R/σ^2. The results of §**38** imply that R_i/σ^2 is distributed as $\chi^2_{n_i-1}$ independently of M_i. Consequently R/σ^2 is distributed as χ^2_{n-k} independently of Q/σ^2.

Let B_1 and B_2 be independent variates where B_1 is $\chi^2_{\nu_1}$ and B_2 is $\chi^2_{\nu_2}$. Define

$$F_{\nu_1, \nu_2} = (B_1/\nu_1)/(B_2/\nu_2).$$

The density function of F_{ν_1,ν_2} can be found by the methods of §**22** (see exercise 5). We say that F_{ν_1,ν_2} has Fisher's F-distribution with ν_1 degrees of freedom in the numerator and ν_2 degrees of freedom in the denominator. Denote the upper 100ε percent point by $F_{\nu_1,\nu_2;\,\varepsilon}$. A table of percentage points is given by Pearson and Hartley (1966).

By taking $B_1 = Q/\sigma^2$, $B_2 = R/\sigma^2$, $\nu_1 = k-1$, and $\nu_2 = n-k$, we find that the variance-ratio is distributed as $F_{k-1,n-k}$ when $\mathscr{H}$ is true. A test procedure at significance level ε therefore consists in rejecting $\mathscr{H}$ whenever

$$\{q/(k-1)\}/\{r/(n-k)\} \geqslant F_{k-1,n-k;\,\varepsilon}.$$

§41. Analysis of variance

Consider the identity

$$(X_{ij}-M) \equiv (M_i-M)+(X_{ij}-M_i).$$

Square both sides and sum over i and j. The product term vanishes because

$$\sum_{i=1}^{k}\sum_{j=1}^{n_i}(M_i-M)(X_{ij}-M_i) = \sum_{i=1}^{k}(M_i-M)\left\{\sum_{j=1}^{n_i}(X_{ij}-M_i)\right\},$$

and the last sum on the right is zero. Hence

$$\begin{aligned}\sum_{i=1}^{k}\sum_{j=1}^{n_i}(X_{ij}-M)^2 &\equiv \sum_{i=1}^{k} n_i(M_i-M)^2+\sum_{i=1}^{k}\sum_{j=1}^{n_i}(X_{ij}-M_i)^2\\ &= Q+R.\end{aligned}$$

Each term in this identity is associated with a different source of variation and can be used to estimate σ^2 when $\mathscr{H}$ is true. Q measures the variation between sample means, R measures the variation

TABLE 9

Analysis of variance between and within samples

Source of variation	*Sums of squares*	*Degrees of freedom*	*Mean squares*	*Expectations of mean squares*
Between samples	$\Sigma n_i(M_i-M)^2$	$k-1$	$Q/(k-1)$	$\sigma^2+\Sigma n_i(\mu_i-\mu)^2/(k-1)$
Within samples	$\Sigma\Sigma(X_{ij}-M_i)^2$	$n-k$	$R/(n-k)$	σ^2
Total	$\Sigma\Sigma(X_{ij}-M)^2$	$n-1$		

within samples, and $\Sigma\Sigma(X_{ij}-M)^2$ measures the total variation of the sample values about the overall mean. All are described as sums of squares. The set of statistical methods which are associated with such identities is known as **analysis of variance**. Table 9 is a summary of the main results for an analysis of variance between and within samples. Much the same tabular form is used for the numerical counterpart.

The calculations proceed as follows. Write

$$b_i = \sum_{j=1}^{n_i} x_{ij},\ b = \sum_{i=1}^{k} b_i,\ c_i = \sum_{j=1}^{n_i} x_{ij}^2,\ c = \sum_{i=1}^{k} c_i.$$

Thus b_i is the total of the ith sample, b is the grand total, and c is the sum of all the squared observations. The sum of squares between samples is

$$\sum_{i=1}^{k} n_i m_i^2 - nm^2 = \sum_{i=1}^{k} b_i^2/n_i - b^2/n,$$

the sum of squares within samples is

$$\sum_{i=1}^{k}\sum_{j=1}^{n_i} (x_{ij}-m_i)^2 = c - \sum_{i=1}^{k} b_i^2/n_i,$$

and the total sum of squares is

$$\sum_{i=1}^{k}\sum_{j=1}^{n_i} (x_{ij}-m)^2 = c - b^2/n.$$

TABLE 10

Effects of relaxin on interpubic gap in four groups of mice

Group	*Increase in gap (mm.)*	n_i	b_i	c_i
1	0·15, 0·50, 0·40, 0·40, 0·30	5	1·75	0·6825
2	1·90, 2·30, 1·35, 1·50, 1·40, 1·20	6	9·65	16·3725
3	2·00, 2·20, 1·20, 1·40, 2·20, 0, 0·50	7	9·50	17·3300
4	1·50, 2·50, 2·50, 1·50	4	8·00	17·0000

Source of variation	*Sums of squares*	*Degrees of freedom*	*Mean squares*	*Variance ratio*
Between groups	7·0610	3	2·3537	6·66
Within groups	6·3591	18	0·3533	—
Total	13·4201	21		

Each quantity is computed from the expression on the right, and we therefore require $\Sigma b_i^2/n_i$, b^2/n, and c.

Example. In one of the series of experiments described by Hall (1948), four groups of mice were given different numbers of daily injections of oestrone. A measurement was made on each mouse of the increase in the interpubic gap 24 hours after a single injection of relaxin. The measurements, the basic statistics, and the analysis of variance are given in Table 10. On comparing the observed variance-ratio of 6·66 with $F_{3,18;0\cdot01} = 5\cdot09$, we infer that there are differences between $\mu_1, \ldots, \mu_4$. A further discussion of the results in relation to specified objectives is made in the reference cited.

§42. Fixed and random effects

We have discussed estimation and test procedures for k normal samples on the basis of what is called a **fixed-effects** model, in which the parameters $\mu_1, \mu_2, \ldots, \mu_k$ have constant values. However, situations occur where the assumption that $\mu_1, \mu_2, \ldots, \mu_k$ are fixed is inappropriate. For example, consider a factory containing a large number of identical machines which each manufacture steel rods. A selection of k machines is made, a sample of n_i rods is taken from the ith machine selected, and the diameter of each rod is measured. Here x_{ij} is the diameter of the jth rod from the ith machine. Thus μ_i, which represents the average diameter for the ith machine, is not fixed because of the selection process involved. In situations of this type, a **random-effects** model is introduced. A convenient version is the following. We assume that

$$X_{ij} = \mu_i + e_{ij},$$

where the $\{\mu_i\}$ and $\{e_{ij}\}$ are mutually independent normal variates, such that the $\{\mu_i\}$ are identically distributed with mean ξ and variance σ_1^2, and the $\{e_{ij}\}$ are identically distributed with mean zero and variance σ^2. A consequence of the assumptions is that

$$M_i = \mu_i + \sum_{j=1}^{n_i} e_{ij}/n_i,$$

whence $E(M_i) = \xi$ and $\text{var}\,(M_i) = \sigma_1^2 + \sigma^2/n_i$.

We begin the analysis by discussing the estimation of σ^2 and σ_1^2. Consider the sums of squares in the analysis of variance for the fixed-effects model. The expectation of Q conditional on fixed $\{\mu_i\}$ is

$$E(Q \mid \mu_1, \mu_2, \ldots, \mu_k) = (k-1)\sigma^2 + \Sigma n_i(\mu_i - \mu)^2.$$

For any ξ, this can be written in the form

$$E(Q \mid \mu_1, \mu_2, \ldots, \mu_k) = (k-1)\sigma^2 + \Sigma n_i(\mu_i - \xi)^2 - n(\mu - \xi)^2.$$

Now var $(\mu_i) = \sigma_1^2$ and var $(\mu) = \sigma_1^2 \Sigma n_i^2/n^2$. The unconditional expectation of Q is therefore

$$E(Q) = (k-1)\sigma^2 + \sigma_1^2(n^2 - \Sigma n_i^2)/n.$$

We say that a **balanced** experimental design is used if

$$n_1 = n_2 = \ldots = n_k.$$

Denote the common value by a, in which case

$$E(Q) = (k-1)(\sigma^2 + a\sigma_1^2).$$

The expectation of R conditional on fixed $\{\mu_i\}$ does not involve the $\{\mu_i\}$, and hence the unconditional expectation is

$$E(R) = (n-k)\sigma^2.$$

Unbiased estimators of σ^2 and σ_1^2 are given by the values of these parameters which satisfy the equations

$$Q = E(Q) \quad \text{and} \quad R = E(R).$$

Denote the estimates by s^2 and s_1^2 respectively. We find that

$$s^2 = r/(n-k) \quad \text{and} \quad s_1^2 = n\{q - (k-1)r/(n-k)\}/(n^2 - \Sigma n_i^2).$$

A negative solution for s_1^2 is possible. If we adopt the natural procedure of recording an estimate of zero when $s_1^2 < 0$, then the estimator becomes biased.

We now derive a test of $\sigma_1^2 = 0$. In order to simplify the distribution theory, we assume in what follows that the design is balanced. The variates $\{M_i\}$ are normally and independently distributed, each with mean ξ and variance $\sigma_1^2 + \sigma^2/a$. Thus $\Sigma(M_i - M)^2/(k-1)$ is distributed as the variance of a sample of size k from $\mathcal{N}(\xi, \sigma_1^2 + \sigma^2/a)$. Consequently $\Sigma(M_i - M)^2/(\sigma_1^2 + \sigma^2/a)$ is χ_{k-1}^2, and since $Q = a\Sigma(M_i - M)^2$, we find that Q is distributed as $(\sigma^2 + a\sigma_1^2)\chi_{k-1}^2$. The distribution of R conditional on fixed $\{\mu_i\}$ does not involve the $\{\mu_i\}$ and hence the unconditional distribution is $\sigma^2\chi_{n-k}^2$. Since Q and R are independent for any fixed $\{\mu_i\}$, their joint density function conditional on fixed $\{\mu_i\}$ can be factorized in the form

$$h(q, r \mid \mu_1, \mu_2, \ldots, \mu_k) = f(q \mid \mu_1, \mu_2, \ldots, \mu_k)g(r).$$

Integration over $\mu_1, \mu_2, \ldots, \mu_k$ shows that Q and R are independent unconditionally. We conclude that the variance-ratio is distributed as $(1+a\sigma_1^2/\sigma^2)F_{k-1,n-k}$. Thus a test of $\sigma_1^2 = 0$ can be made by referring the variance-ratio to a table of $F_{k-1,n-k;\,\varepsilon}$, exactly as in the case of fixed effects.

EXERCISES 8

1. A sample of size n is taken from $\mathcal{N}(\mu, \sigma^2)$ and the standard deviation is S. Prove that

$$E\left[S\left(\frac{n-1}{2}\right)^{\frac{1}{2}} \frac{\Gamma\{\frac{1}{2}(n-1)\}}{\Gamma(\frac{1}{2}n)}\right] = \sigma.$$

Let ε_1 and ε_2 be probabilities such that $\varepsilon_1+\varepsilon_2 = \varepsilon$. Show that the interval from $(n-1)^{\frac{1}{2}}s/\chi_{n-1;\,\varepsilon_2}$ to $(n-1)^{\frac{1}{2}}s/\chi_{n-1;\,1-\varepsilon_1}$ is a confidence interval for σ with confidence coefficient $1-\varepsilon$.

2. A sample of size n is taken from $\mathcal{N}(0, 1)$. Denote the range by V_n and its upper 100ε percent point by $v_{n;\,\varepsilon}$. Let ε_1 and ε_2 be probabilities such that $\varepsilon_1+\varepsilon_2 = \varepsilon$. Denote by W_n the range in a sample of size n from $\mathcal{N}(\mu, \sigma^2)$. Show that $(w_n/v_{n;\,\varepsilon_2}, w_n/v_{n;\,1-\varepsilon_1})$ is a confidence interval for σ with confidence coefficient $1-\varepsilon$.

3. $X_1, X_2, \ldots, X_n$ are mutually independent standard normal variates. Let $k(x_1, x_2, \ldots, x_n)$ be a function with the property

$$k(x_1+a, x_2+a, \ldots, x_n+a) = k(x_1, x_2, \ldots, x_n) \quad \text{for all } a.$$

Find the joint moment generating function of ΣX_i and $k(X_1, X_2, \ldots, X_n)$. Hence prove that these variates are independent.

Deduce that the mean and range are independent in samples from $\mathcal{N}(\mu, \sigma^2)$.

4. A and B are independent variates such that A is $\mathcal{N}(0, 1)$ and B is χ_ν^2. Define $T = A/(B/\nu)^{\frac{1}{2}}$. Show that the density function of T is given by

$$\frac{\Gamma\{\frac{1}{2}(\nu+1)\}}{\Gamma(\frac{1}{2}\nu)(\pi\nu)^{\frac{1}{2}}}\left(1+\frac{t^2}{\nu}\right)^{-\frac{1}{2}(\nu+1)}.$$

Prove that T^2 is distributed as $F_{1,\nu}$.

5. B_1 and B_2 are independent variates which are distributed as $\chi_{\nu_1}^2$ and $\chi_{\nu_2}^2$ respectively. Define $F_{\nu_1,\nu_2} = (B_1/\nu_1)/(B_2/\nu_2)$. Show that the density function of F_{ν_1,ν_2} is given by

$$[\Gamma\{\tfrac{1}{2}(\nu_1+\nu_2)\}/\Gamma(\tfrac{1}{2}\nu_1)\Gamma(\tfrac{1}{2}\nu_2)](\nu_1 f/\nu_2)^{\frac{1}{2}\nu_1-1}(1+\nu_1 f/\nu_2)^{-\frac{1}{2}(\nu_1+\nu_2)}(\nu_1/\nu_2).$$

Prove that $F_{\nu_2,\nu_1;\,\varepsilon} = 1/F_{\nu_1,\nu_2;\,1-\varepsilon}$.

6. Show that the variance-ratio for testing the hypothesis $\mu_1 = \mu_2$ can be expressed as

$$(M_1-M_2)^2\bigg/\left\{\frac{(n_1-1)S_1^2+(n_2-1)S_2^2}{n_1+n_2-2}\left(\frac{1}{n_1}+\frac{1}{n_2}\right)\right\}.$$

Deduce that the test of $\mu_1 = \mu_2$ in §**39** is a special case of the test of $\mathscr{H}$ in §**40**.

7. B_1 and B_2 are independent variates which are distributed as $\chi^2_{\nu_1}$ and $\chi^2_{\nu_2}$ respectively. Define $G_{\nu_1,\nu_2} = B_1/(B_1+B_2)$. Prove that G_{ν_1,ν_2} is distributed independently of B_1+B_2 with density function

$$[\Gamma\{\tfrac{1}{2}(\nu_1+\nu_2)\}/\Gamma(\tfrac{1}{2}\nu_1)\Gamma(\tfrac{1}{2}\nu_2)]g^{\frac{1}{2}\nu_1-1}(1-g)^{\frac{1}{2}\nu_2-1} \quad (0 < g < 1).$$

Obtain a connection between the percentage points of F_{ν_1,ν_2} and those of G_{ν_1,ν_2}.

8. Samples of sizes n_1 and n_2 are taken from $\mathscr{N}(\mu_1, \sigma_1^2)$ and $\mathscr{N}(\mu_2, \sigma_2^2)$ respectively. The sample variances are S_1^2 and S_2^2. We require to test the hypothesis $\mathscr{H}: \sigma_1^2 = \sigma_2^2$ when the alternative is $\sigma_1^2 > \sigma_2^2$. Write $\theta = \sigma_1^2/\sigma_2^2$. Show that a test at significance level ε is obtained by rejecting $\mathscr{H}$ whenever $s_1^2/s_2^2 > F_{n_1-1,n_2-1;\,\varepsilon}$. Prove that the power function of the test is given by

$$p(\mathscr{H} \text{ rejected} \mid \sigma_1^2, \sigma_2^2) = p(F_{n_1-1,n_2-1} > F_{n_1-1,n_2-1;\,\varepsilon}/\theta).$$

This test is not recommended because the power function is seriously affected by departures from normality.

9. The experiments on the density of nitrogen described in the example of §**38** also gave the following results from using ferrous hydrate as a reducing agent.

2·31024 2·31010 2·31028

Test the hypothesis that the mean weight of nitrogen in the bulb is the same for each of the three methods.

10. Samples of sizes $n_1, n_2, \ldots, n_k$ are taken from $\mathscr{N}(\mu_1, \sigma^2)$, $\mathscr{N}(\mu_2, \sigma^2), \ldots, \mathscr{N}(\mu_k, \sigma^2)$ respectively. Define

$$D = \Sigma n_i(M_i-\mu_i)^2 - n(M-\mu)^2.$$

Show that D/σ^2 is χ^2_{k-1} and independent of R/σ^2.

Consider the region in k-dimensional space defined by all points $(u_1, u_2, \ldots, u_k)$ which satisfy

$$(n-k)\{\Sigma n_i(M_i-u_i)^2-n(M-u)^2\} \leqslant R(k-1)F_{k-1,n-k;\,\varepsilon},$$

where $u = \Sigma n_i u_i/n$. Prove that this region contains with probability $1-\varepsilon$ the point $(\mu_1, \mu_2, \ldots, \mu_k)$. We say that the region observed is a **confidence region** for $(\mu_1, \mu_2, \ldots, \mu_k)$ with confidence coefficient $1-\varepsilon$.

CHAPTER 9
LINEAR STATISTICAL MODELS

§43. Linear regression

Consider an experiment where the data consist of n pairs of measurements, denoted by $(x_1, y_1), (x_2, y_2), \ldots, (x_n, y_n)$. There are two statistical models which may be relevant in practice, and which lead to the same method of analysis.

According to the first model, $(x_1, y_1), (x_2, y_2), \ldots, (x_n, y_n)$ form a random sample of size n from the bivariate distribution of X and Y, where the regression of Y on X is linear and the distribution of Y conditional on $X = x$ has a variance independent of x. By the definition of a random sample from a bivariate distribution, the data consist of observations on $(X_1, Y_1), (X_2, Y_2), \ldots, (X_n, Y_n)$ respectively, where the pairs of variates each have the same distribution as (X, Y) and are mutually independent. The other assumptions can then be written in the form

$$E(Y_i \mid x_i) = \theta + \beta x_i \quad \text{and} \quad \operatorname{var}(Y_i \mid x_i) = \sigma^2,$$

for $i = 1, 2, \ldots, n$.

Example. In experiments described by Wallace (1954), larvae of a wood-boring beetle were fed on wood in various stages of decay, and the conditions of temperature and relative humidity were maintained at a constant level. The results for slightly decayed wood are reproduced in Table 11, which gives the initial weight of the larvae (x) and the weight after 35 days (y). Both x and y are expressed in mgm. An appropriate assumption here is that the 23 pairs of observations form a random sample from a bivariate distribution where X is initial weight and Y is final weight. Each pair (x, y) is represented by a point in Fig. 13, and the tendency for y to increase linearly with

TABLE 11

Larvae fed on slightly decayed wood: initial weight (x) and weight after 35 days (y), both in mgm.

x	y	x	y	x	y	x	y	x	y
74	85	98	107	132	165	149	189	161	196
77	91	112	134	135	166	150	182	163	195
79	95	120	146	136	169	153	180	197	231
84	104	128	169	147	172	159	195		
95	110	129	156	148	173	161	186		

x suggests that the regression of Y on X is linear. There is no information about var $(Y \mid x)$, but on the principle that the model should be as simple as possible we assume that var $(Y \mid x)$ is independent of x.

According to the second model, $y_1, y_2, \ldots, y_n$ are observations on mutually independent variates $Y_1, Y_2, \ldots, Y_n$ respectively, whereas $x_1, x_2, \ldots, x_n$ are variables which can be assigned definite values by the experimenter and are not connected with any probability distribution. Furthermore,

$$E(Y_i) = \theta + \beta x_i \quad \text{and} \quad \text{var}(Y_i) = \sigma^2$$

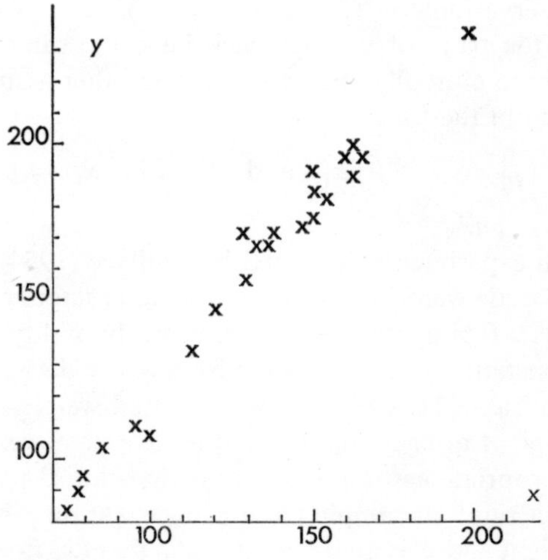

Fig. 13. Relation between final weight of larvae and initial weight.

for $i = 1, 2, \ldots, n$. This model is applicable much less frequently than the first one.

For each model, θ, β and σ^2 are parameters which have fixed but unknown values, and the problem arises of estimating them from the pairs of measurements. Further assumptions about the distribution of $Y_1, Y_2, \ldots, Y_n$ are necessary before we can construct confidence intervals and significance tests for these parameters and other quantities of interest such as $\theta+\beta x$. When the first model is appropriate, the assumptions usually made refer to the distribution of $Y_1, Y_2, \ldots, Y_n$ conditional on $X_1, X_2, \ldots, X_n$ taking their observed values of $x_1, x_2, \ldots, x_n$ respectively, and any conclusions we draw are naturally subject to the same conditions. The terminology of the first model is customarily extended to the somewhat different circumstances of the second, and both are described as linear regression models. They have been extensively studied and generalized. The totality of methods which has resulted is known as **regression analysis.**

§44. Principle of least squares

We now discuss the analysis of experimental data consisting of n pairs of measurements $(x_1, y_1), (x_2, y_2), \ldots, (x_n, y_n)$. The assumptions are as follows.

(1) $x_1, x_2, \ldots, x_n$ are not all equal.
(2) $y_1, y_2, \ldots, y_n$ are observations on mutually independent variates $Y_1, Y_2, \ldots, Y_n$ respectively.
(3) $E(Y_i) = \theta+\beta x_i$ and $\text{var}(Y_i) = \sigma^2 \quad (i = 1, 2, \ldots, n)$.

Denote $\Sigma x_i/n$ by $\bar{x}$ and $\theta+\beta\bar{x}$ by α. Thus assumption (3) can be written as

$$E(Y_i) = \alpha+\beta(x_i-\bar{x}) \quad \text{and} \quad \text{var}(Y_i) = \sigma^2 \quad (i = 1, 2, \ldots, n).$$

The change is made simply as a matter of convenience, for reasons given later.

We turn first to the estimation of α and β. Define

$$w = \sum_{i=1}^{n} \{y_i-\alpha-\beta(x_i-\bar{x})\}^2.$$

As shown in Fig. 14, $y_i-\alpha-\beta(x_i-\bar{x})$ is the deviation of the point (x_i, y_i) from the line $y = \alpha+\beta(x-\bar{x})$, measured parallel to the y-axis. Thus w is the sum of squared deviations, and it measures how close to the line $y = \alpha+\beta(x-\bar{x})$ is the set of

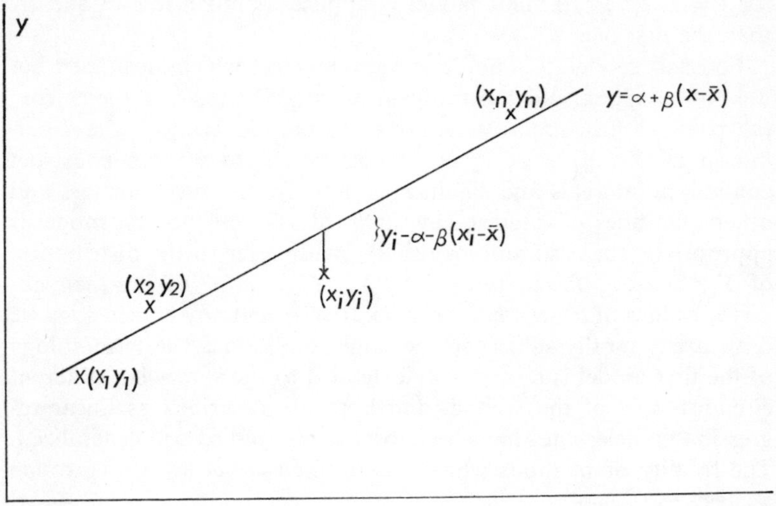

Fig. 14. Deviation of (x_i, y_i) from $y = \alpha + \beta(x - \bar{x})$ measure parallel to y-axis.

points (x_i, y_i). Now consider α and β as variables, and suppose that the absolute minimum of w is attained at $\alpha = a$ and $\beta = b$, where a and b are functions of $x_1, y_1, \ldots, x_n, y_n$. The estimation of the fixed but unknown parameters α and β by a and b respectively is intuitively acceptable because the line $y = a + b(x - \bar{x})$ is then closest to the points (x_i, y_i) in the sense defined.

We say that the procedure of selecting estimates by minimizing w is obtained from the **principle of least squares**, and we describe a and b as least squares estimates. At the minimum of w,

$$\frac{\partial w}{\partial \alpha} = 0 \quad \text{and} \quad \frac{\partial w}{\partial \beta} = 0,$$

where

$$\frac{\partial w}{\partial \alpha} = -2 \sum_{i=1}^{n} \{y_i - \alpha - \beta(x - \bar{x})\},$$

$$\frac{\partial w}{\partial \beta} = -2 \sum_{i=1}^{n} (x_i - \bar{x})\{y_i - \alpha - \beta(x_i - \bar{x})\}.$$

Since $\Sigma(x_i-\bar{x}) = 0$, the estimates satisfy the equations

$$na = \sum_{i=1}^{n} y_i \quad \text{and} \quad b \sum_{i=1}^{n} (x_i-\bar{x})^2 = \sum_{i=1}^{n} (x_i-\bar{x})y_i.$$

From assumption (1), $\Sigma(x_i-\bar{x})^2 > 0$. The estimates are therefore

$$a = \bar{y} \quad \text{and} \quad b = \sum_{i=1}^{n} (x_i-\bar{x})y_i \Big/ \sum_{i=1}^{n} (x_i-\bar{x})^2,$$

where $\bar{y}$ stands for $\Sigma y_i/n$.

The argument is completed by showing that w attains its absolute minimum when $\alpha = a$ and $\beta = b$. This can be proved by further differentiation, or as follows. Consider the identity

$$\{y_i-\alpha-\beta(x_i-\bar{x})\} \equiv (a-\alpha)+(x_i-\bar{x})(b-\beta)+\{y_i-a-b(x_i-\bar{x})\}.$$

Square both sides and sum over $i = 1, 2, \ldots, n$. The product terms are all zero. For example,

$$(b-\beta) \sum_{i=1}^{n} (x_i-\bar{x})\{y_i-a-b(x_i-\bar{x})\} = 0$$

because the sum is $-\frac{1}{2}\partial w/\partial\beta$ evaluated at $\alpha = a$, $\beta = b$. The other products vanish because $\Sigma(x_i-\bar{x}) = 0$ and $\Sigma(y_i-\bar{y}) = 0$. In this way we obtain the basic identity

$$w \equiv n(a-\alpha)^2+(b-\beta)^2 \sum_{i=1}^{n} (x_i-\bar{x})^2+ \sum_{i=1}^{n} \{y_i-a-b(x_i-\bar{x})\}^2.$$

All the terms on the right are non-negative. The first is minimized when $\alpha = a$, the second is minimized when $\beta = b$, and the third is independent of α and β. Therefore $\alpha = a$, $\beta = b$ gives an absolute minimum of w.

Corresponding to the estimates a and b are the estimators

$$\hat{\alpha} = \sum_{i=1}^{n} Y_i/n \quad \text{and} \quad \hat{\beta} = \sum_{i=1}^{n} (x_i-\bar{x})\, Y_i \Big/ \sum_{i=1}^{n} (x_i-\bar{x})^2.$$

We note first that $\hat{\alpha}$ and $\hat{\beta}$ are unbiased estimators of α and β respectively. The argument depends on the fact that $\Sigma(x_i-\bar{x}) = 0$, whence

$$E(\hat{\alpha}) = \sum_{i=1}^{n} \{\alpha+\beta(x_i-\bar{x})\}/n = \alpha,$$

and

$$E(\hat{\beta}) = \sum_{i=1}^{n} (x_i - \bar{x})\{\alpha + \beta(x_i - \bar{x})\} \,/\, \sum_{i=1}^{n} (x_i - \bar{x})^2 = \beta.$$

Now consider the variances and covariance of $\hat{\alpha}$ and $\hat{\beta}$. The variance of a linear combination of variates was derived at the end of §**19**, and the covariance of two such combinations is found by similar arguments. We thus obtain

$$\text{var}\,(\hat{\alpha}) = \sum_{i=1}^{n} \text{var}\,(Y_i)/n^2 = \sigma^2/n,$$

$$\text{var}\,(\hat{\beta}) = \sum_{i=1}^{n} (x_i - \bar{x})^2 \,\text{var}\,(Y_i)/\{\sum_{i=1}^{n} (x_i - \bar{x})^2\}^2$$

$$= \sigma^2 \,/\, \sum_{i=1}^{n} (x_i - \bar{x})^2,$$

and

$$\text{cov}\,(\hat{\alpha}, \hat{\beta}) = \sum_{i=1}^{n} (x_i - \bar{x})\,\text{var}\,(Y_i) \,/\, n \sum_{i=1}^{n} (x_i - \bar{x})^2 = 0.$$

We say that the estimators $\hat{\alpha}$ and $\hat{\beta}$ are linear because each is a linear combination of $Y_1, Y_2, \ldots, Y_n$, and we have shown that $\hat{\alpha}$ and $\hat{\beta}$ are unbiased. In fact, $\hat{\alpha}$ is the best linear unbiased estimator of α in the sense that the variance of any other linear unbiased estimator of α exceeds var ($\hat{\alpha}$). Similarly, $\hat{\beta}$ is the best linear unbiased estimator of β. We shall prove the result for $\hat{\beta}$ and note that the proof for $\hat{\alpha}$ follows identical lines. Suppose that $\Sigma c_i Y_i$ is an unbiased estimator of β. Then

$$\sum_{i=1}^{n} c_i\{\alpha + \beta(x_i - \bar{x})\} = \beta \quad \text{for all } \alpha \text{ and } \beta,$$

so that the coefficients $\{c_i\}$ must satisfy the conditions

$$\sum_{i=1}^{n} c_i = 0 \quad \text{and} \quad \sum_{i=1}^{n} c_i(x_i - \bar{x}) = 1.$$

The least squares estimator $\hat{\beta}$ has coefficients $\{(x_i - \bar{x})/\Sigma(x_i - \bar{x})^2\}$. Consider

$$\Sigma\left\{c_i - \frac{x_i - \bar{x}}{\Sigma(x_i - \bar{x})^2}\right\}^2 = \Sigma c_i^2 - \frac{2\Sigma c_i(x_i - \bar{x})}{\Sigma(x_i - \bar{x})^2} + \frac{1}{\Sigma(x_i - \bar{x})^2}.$$

The condition $\Sigma c_i(x_i - \bar{x}) = 1$ implies that

$$\Sigma\left\{c_i - \frac{x_i - \bar{x}}{\Sigma(x_i - \bar{x})^2}\right\}^2 = \Sigma c_i^2 - \frac{1}{\Sigma(x_i - \bar{x})^2}.$$

Now var $(\Sigma c_i Y_i) = \sigma^2 \Sigma c_i^2$, and so we obtain

$$\text{var}\,(\Sigma c_i Y_i) = \frac{\sigma^2}{\Sigma(x_i - \bar{x})^2} + \sigma^2 \Sigma\left\{c_i - \frac{x_i - \bar{x}}{\Sigma(x_i - \bar{x})^2}\right\}^2.$$

The second term on the right is non-negative. Therefore var $(\Sigma c_i Y_i)$ is minimized when $c_i = (x_i - x)/\Sigma(x_i - \bar{x})^2$, and the coefficients then satisfy the condition that $\Sigma c_i = 0$. This shows that $\hat{\beta}$ is the best linear unbiased estimator of β.

The deviation of the point (x_i, y_i) from the line $y = a + b(x - \bar{x})$, measured parallel to the y-axis, is $y_i - a - b(x_i - \bar{x})$. This is called a **residual**. Define

$$r = \sum_{i=1}^{n} \{y_i - a - b(x_i - \bar{x})\}^2.$$

We say that r is the **sum of squared residuals**. It is the value of w when $\alpha = a$, $\beta = b$, and therefore r is the absolute minimum of w. Corresponding to w and r are the variates

$$W = \sum_{i=1}^{n} \{Y_i - \alpha - \beta(x_i - \bar{x})\}^2 \text{ and } R = \sum_{i=1}^{n} \{Y_i - \hat{\alpha} - \hat{\beta}(x_i - \bar{x})\}^2.$$

In terms of variates, the basic identity is

$$W \equiv n(\hat{\alpha} - \alpha)^2 + (\hat{\beta} - \beta)^2 \sum (x_i - \bar{x})^2 + R.$$

Taking expectations throughout, we find that

$$\Sigma\,\text{var}\,(Y_i) = n\,\text{var}\,(\hat{\alpha}) + \text{var}\,(\hat{\beta})\Sigma(x_i - \bar{x})^2 + E(R).$$

On substituting for var $(\hat{\alpha})$ and var $(\hat{\beta})$, we obtain

$$E(R) = (n-2)\sigma^2.$$

Thus $R/(n-2)$ is an unbiased estimator of σ^2.

Suppose that we want to test a hypothesis $\mathscr{H}$ which assigns a specific value to some linear combination of α and β. Let R_0 be the relative minimum of W when $\mathscr{H}$ is true. The sum of squares for testing $\mathscr{H}$ is denoted by Q and defined by

$$Q = R_0 - R.$$

This definition accords with the one given in §**39**. Consider first the hypothesis that $\beta = 0$. When $\beta = 0$ is inserted in the basic identity, the expression which results is minimized when $\alpha = \hat{\alpha}$, and the minimum value is $\hat{\beta}^2\Sigma(x_i - \bar{x})^2 + R$. Let Q_β be the sum of squares for testing $\beta = 0$. Then

$$Q_\beta = \hat{\beta}^2\Sigma(x_i - \bar{x})^2.$$

In the same way, we deduce that the sum of squares for testing $\alpha = 0$, say Q_α, is given by

$$Q_\alpha = n\hat{\alpha}^2.$$

Put $\alpha = \beta = 0$ in the basic identity, whence

$$\Sigma Y_i^2 \equiv Q_\alpha + Q_\beta + R.$$

Now the expectation of Q_β is

$$\Sigma(x_i - \bar{x})^2\{\text{var}(\hat{\beta}) + \beta^2\} = \sigma^2 + \beta^2\Sigma(x_i - \bar{x})^2.$$

Therefore an appropriate statistic for testing $\beta = 0$ is the variance-ratio $Q_\beta/\{R/(n-2)\}$, for reasons which parallel those motivating the test of §**40**. Similar remarks apply to Q_α. We say that Q_α measures the variation due to the mean, Q_β that due to regression, and R measures the residual variation. Table 12 summarizes the main

TABLE 12

Analysis of variance for linear regression

Source of variation	*Sums of squares*	*Degrees of freedom*	*Mean squares*	*Expectations of mean squares*
Mean	Q_α	1	Q_α	$\sigma^2 + n\alpha^2$
Regression	Q_β	1	Q_β	$\sigma^2 + \beta^2\Sigma(x_i - \bar{x})^2$
Residual	R	$n-2$	$R/(n-2)$	σ^2
Total	ΣY_i^2	n		

results in the analysis of variance for linear regression. The values given for degrees of freedom anticipate the distribution theory of the next section.

§45. Distributions of linear regression

The results of the previous section are valid whenever the initial assumptions are satisfied, and these specify only the means and variances of the mutually independent variates $Y_1, Y_2, \ldots, Y_n$. We now make the further assumption that $Y_1, Y_2, \ldots, Y_n$ are normal variates. The likelihood function is

$$l(\alpha, \beta, \sigma^2; y_1, y_2, \ldots, y_n) = (2\pi\sigma^2)^{-\frac{1}{2}n} \exp(-w/2\sigma^2)$$

and $$\log l = -\tfrac{1}{2}n \log(2\pi\sigma^2) - w/2\sigma^2.$$

Thus the maximum likelihood estimates of α and β are the values which minimize w. The assumption of normality therefore implies that the principle of least squares is a consequence of the principle of maximum likelihood. We proceed to find the distributions of the estimators and of other relevant quantities.

Denote $\{Y_i - \alpha - \beta(x_i - \bar{x})\}/\sigma$ by Z_i, so that $Z_1, Z_2, \ldots, Z_n$ are mutually independent $\mathcal{N}(0, 1)$ variates. Consider an orthonormal linear transformation from $Z_1, Z_2, \ldots, Z_n$ to $T_1, T_2, \ldots, T_n$ such that

$$T_1 = n^{-\frac{1}{2}} \sum_{i=1}^{n} Z_i \quad \text{and} \quad T_2 = \sum_{i=1}^{n} (x_i - \bar{x}) Z_i \Big/ \Big\{ \sum_{i=1}^{n} (x_i - \bar{x})^2 \Big\}^{\frac{1}{2}}.$$

The coefficients in T_1 and T_2 define orthonormal vectors, because

$$\sum_{i=1}^{n} n^{-1} = 1, \quad \sum_{i=1}^{n} \left[(x_i - \bar{x}) \Big/ \Big\{ \sum_{i=1}^{n} (x_i - \bar{x})^2 \Big\}^{\frac{1}{2}} \right]^2 = 1,$$

and

$$\sum_{i=1}^{n} \left[(x_i - \bar{x}) \Big/ \Big\{ n \sum_{i=1}^{n} (x_i - \bar{x})^2 \Big\}^{\frac{1}{2}} \right] = 0.$$

Substituting for Z_i in T_1 and T_2, we obtain

$$T_1 = n^{-\frac{1}{2}} \sum_{i=1}^{n} \{Y_i - \alpha - \beta(x_i - \bar{x})\}/\sigma = n^{\frac{1}{2}}(\hat{\alpha} - \alpha)/\sigma,$$

and

$$T_2 = \sum_{i=1}^{n} (x_i - \bar{x})\{Y_i - \alpha - \beta(x_i - \bar{x})\}/\sigma \Big\{ \sum_{i=1}^{n} (x_i - \bar{x})^2 \Big\}^{\frac{1}{2}}$$

$$= (\hat{\beta} - \beta) \Big\{ \sum_{i=1}^{n} (x_i - \bar{x})^2 \Big\}^{\frac{1}{2}} \Big/ \sigma.$$

The basic identity can be written as

$$W/\sigma^2 = n(\hat{\alpha} - \alpha)^2/\sigma^2 + (\hat{\beta} - \beta)^2 \sum (x_i - \bar{x})^2/\sigma^2 + R/\sigma^2.$$

By definition, $W/\sigma^2 = \Sigma_1^n Z_i^2$, and by orthonormality, $\Sigma_1^n Z_i^2 = \Sigma_1^n T_j^2$. Thus

$$R/\sigma^2 = \sum_{j=3}^{n} T_j^2.$$

We have shown in example 3 of §**22** that an orthonormal linear transformation of a set of mutually independent $\mathcal{N}(0, 1)$ variates leads to another set of mutually independent $\mathcal{N}(0, 1)$ variates. Consequently $n^{\frac{1}{2}}(\hat{\alpha}-\alpha)/\sigma$ is $\mathcal{N}(0, 1)$, $(\hat{\beta}-\beta)\{\Sigma(x_i-\bar{x})^2\}^{\frac{1}{2}}/\sigma$ is $\mathcal{N}(0, 1)$, R/σ^2 is χ^2_{n-2}, and the three variates are mutually independent. We chose to express $E(Y_i)$ in terms of α and β because $\hat{\alpha}$ and $\hat{\beta}$ are independent.

An immediate consequence of these results is that Q_β/σ^2 is χ_1^2 when $\beta = 0$, and the variance-ratio is then $F_{1,n-2}$. Similar remarks apply in respect of Q_α/σ^2. The results also enable us to construct confidence intervals for any linear combination of α and β. As an illustration, consider $\alpha+\beta(x-\bar{x})$, which is a typical value of $E(Y)$. Denote $\alpha+\beta(x-\bar{x})$ by η and $\hat{\alpha}+\hat{\beta}(x-\bar{x})$ by $\hat{\eta}$. Then $E(\hat{\eta}) = \eta$ and

$$\text{var}(\hat{\eta}) = \sigma^2\left\{\frac{1}{n}+\frac{(x-\bar{x})^2}{\Sigma(x_i-\bar{x})^2}\right\}.$$

Now $\hat{\eta}$ is a linear combination of normal variates and therefore $(\hat{\eta}-\eta)/\{\text{var}(\hat{\eta})\}^{\frac{1}{2}}$ is $\mathcal{N}(0, 1)$. Furthermore, R/σ^2 is independent of $\hat{\eta}$, and so the variate

$$T = \frac{\hat{\eta}-\eta}{\left[\dfrac{R}{n-2}\left\{\dfrac{1}{n}+\dfrac{(x-x)^2}{\Sigma(x_i-\bar{x})^2}\right\}\right]^{\frac{1}{2}}}$$

has 'Student's' t-distribution with $(n-2)$ degrees of freedom. By an argument similar to one used in §**36**, we find that the interval

$$\hat{\eta}\pm t_{n-2;\frac{1}{2}\varepsilon}\left[\frac{R}{n-2}\left\{\frac{1}{n}+\frac{(x-\bar{x})^2}{\Sigma(x_i-\bar{x})^2}\right\}\right]^{\frac{1}{2}}$$

covers η with probability $1-\varepsilon$. Thus

$$a+b(x-\bar{x})\pm t_{n-2;\frac{1}{2}\varepsilon}\left[\frac{r}{n-2}\left\{\frac{1}{n}+\frac{(x-\bar{x})^2}{\Sigma(x_i-\bar{x})^2}\right\}\right]^{\frac{1}{2}}$$

is a confidence interval for $\alpha+\beta(x-\bar{x})$ with confidence coefficient $1-\varepsilon$.

Standard programs are available for regression analysis by computer. When the computing facilities are limited, the data can be analysed on the following lines. Write

$$f = \Sigma(x_i - \bar{x})^2, \quad g = \Sigma(y_i - \bar{y})^2,$$

and

$$h = \Sigma(x_i - \bar{x})(y_i - \bar{y}).$$

We begin by calculating

$$\Sigma x_i, \ \Sigma y_i, \ \Sigma x_i^2, \ \Sigma y_i^2, \text{ and } \Sigma x_i y_i,$$

in terms of which all the quantities required can be expressed. Thus

$$f = \Sigma x_i^2 - (\Sigma x_i)^2/n, \quad g = \Sigma y_i^2 - (\Sigma y_i)^2/n,$$

and

$$h = \Sigma x_i y_i - (\Sigma x_i)(\Sigma y_i)/n.$$

The analysis of variance for linear regression is based on the identity

$$r \equiv \Sigma y_i^2 - (\Sigma y_i)^2/n - b^2 \Sigma(x_i - \bar{x})^2.$$

On substituting $b = h/f$, we obtain

$$r = g - h^2/f.$$

Example 1. Larvae of a wood-boring beetle were fed on slightly decayed wood. The initial weights and the weights after 35 days are given in Table 11 (p. 146). We require a confidence interval for the expected weight after 35 days when the initial weight is 100 mgm.

$$\Sigma x = 2987, \qquad \Sigma y = 3596, \qquad n = 23,$$
$$\Sigma x^2 = 411569, \qquad \Sigma y^2 = 597588, \qquad \Sigma xy = 495620,$$
$$\bar{x} = 129{\cdot}9 \qquad \bar{y} = 156{\cdot}3,$$
$$f = 23649, \qquad g = 35361, \qquad h = 28609,$$
$$a = 156{\cdot}3, \qquad b = 1{\cdot}2097, \qquad r = 752.$$

For a confidence coefficient of 95 percent, $\varepsilon = 0{\cdot}05$, and $t_{21;0\cdot025} = 2{\cdot}08$. The confidence interval is 120 ± 5 mgm.

No further distribution theory is required for a comparison of the regression lines from two experiments, provided that σ^2 has the same value throughout. The first experiment supplies n_1 pairs of measurements typified by (x_{1i}, y_{1i}). We assume that

$$E(Y_{1i}) = \alpha_1 + \beta_1(x_{1i} - \bar{x}_1) \text{ and } \operatorname{var}(Y_{1i}) = \sigma^2 \ (i = 1, 2, \ldots, n_1),$$

where $\bar{x}_1$ denotes $\Sigma x_{1i}/n_1$. The variates $\hat{\alpha}_1$, $\hat{\beta}_1$ and R_1 are defined by analogy with $\hat{\alpha}$, $\hat{\beta}$ and R respectively. In the same way, the second experiment supplies n_2 pairs (x_{2j}, y_{2j}), such that

$$E(Y_{2j}) = \alpha_2 + \beta_2(x_{2j} - \bar{x}_2) \text{ and } \operatorname{var}(Y_{2j}) = \sigma^2 \ (j = 1, 2, \ldots, n_2),$$

where $\bar{x}_2$ denotes $\Sigma x_{2j}/n_2$. The variates $\hat{\alpha}_2$, $\hat{\beta}_2$, and R_2 correspond respectively to $\hat{\alpha}$, $\hat{\beta}$ and R.

Consider the problem of testing the hypothesis $\mathscr{H}: \beta_1 = \beta_2$. Introduce the variates

$$X = \frac{(\hat{\beta}_1 - \hat{\beta}_2) - (\beta_1 - \beta_2)}{\left\{\sigma^2\left(\frac{1}{f_1} + \frac{1}{f_2}\right)\right\}^{\frac{1}{2}}} \quad \text{and } R = R_1 + R_2,$$

where $f_1 = \Sigma(x_{1i} - \bar{x}_1)^2$ and $f_2 = \Sigma(x_{2j} - \bar{x}_2)^2$. Since $\hat{\beta}_1$, $\hat{\beta}_2$, R_1 and R_2 are mutually independent, X and R are independent. Furthermore, X is $\mathscr{N}(0, 1)$, and R/σ^2 is distributed as the sum of $\chi^2_{n_1-2}$ and χ^2_{n2-2}, namely $\chi^2_{n_1+n_2-4}$. Hence

$$T = \frac{(\hat{\beta}_1 - \hat{\beta}_2) - (\beta_1 - \beta_2)}{\left\{\frac{R_1 + R_2}{n_1 + n_2 - 4}\left(\frac{1}{f_1} + \frac{1}{f_2}\right)\right\}^{\frac{1}{2}}}$$

has 'Student's' t-distribution with $n_1 + n_2 - 4$ degrees of freedom. The test procedure implied by this result consists of rejecting $\mathscr{H}$ whenever $|t| \geqslant t_{n_1+n_1-4;\frac{1}{2}\varepsilon}$, where

$$t = \frac{b_1 - b_2}{\left\{\frac{r_1 + r_2}{n_1 + n_2 - 4}\left(\frac{1}{f_1} + \frac{1}{f_2}\right)\right\}^{\frac{1}{2}}},$$

and b_1, b_2, r_1, r_2 are the observed values of $\hat{\beta}_1$, $\hat{\beta}_2$, R_1, R_2 respectively.

Example 2. The experiments described by Wallace (1954) included one group of 23 larvae fed on slightly decayed wood and another group of 17 fed on very decayed wood. In each case, the initial weights and the weights after 35 days were recorded in mgm. The quantities required for a comparison of growth rates are given below.

Type of wood	n	b	r	f
Slightly decayed	23	1·2097	752	23649
Very decayed	17	1·0106	374	142406

Here t is about 5, which corresponds to a significance level below 0·1 percent, and gives a clear indication that the growth rate with slightly decayed wood is larger than with very decayed wood.

§46. Two-way layouts

Consider an experiment which results in cd measurements $\{y_{ij}\}$ where $i = 1, 2, \ldots, c$ and $j = 1, 2, \ldots, d$. The measurements can be displayed as a rectangular array with c rows and d columns, such that y_{ij} occupies the cell formed by the ith row and jth column. When the data are subject to a twofold classification of this type, we describe the experiment as a two-way layout. For example, the variation in the rate of passage of a bismuth meal through the gastro-intestinal tract of normal rats was studied by measuring the emptying time of the stomach in 5 rats which had 7 control bismuth meals (Malaria Research Unit, 1946). The results are presented in Table 13,

TABLE 13

Emptying time of the stomach, in hours, for 5 rats which had 7 control bismuth meals

Rat no.	*Meal no.*						
	1	2	3	4	5	6	7
1	1·75	3·50	2·25	5·50	3·50	3·50	3·50
2	1·75	1·75	3·50	4·50	3·50	2·75	2·75
3	2·25	2·25	3·50	4·50	3·50	3·50	2·75
4	5·50	6·50	6·50	7·50	9·00	5·50	4·50
5	4·50	5·50	3·50	3·50	2·75	5·50	3·50

where the rows correspond to rats and the columns to meals. Each observation is a multiple of 15 min., but approximates to a measurement on a continuous scale for which a linear model of the following kind may be appropriate.

We assume that the variates $\{Y_{ij}\}$ corresponding to $\{y_{ij}\}$ are normally and independently distributed, such that

$$E(Y_{ij}) = \mu+\rho_i+\kappa_j \quad \text{and} \quad \text{var}\,(Y_{ij}) = \sigma^2.$$

Here ρ_i denotes the contribution of the ith row and κ_j that of the jth column. For any values of ρ and κ, we have

$$\begin{aligned}\mu+\rho_i+\kappa_j &= \mu+\rho+\kappa+(\rho_i-\rho)+\kappa_j-\kappa \\ &= \mu'+\rho_i'+\kappa_j', \text{ say.}\end{aligned}$$

This shows that there is an arbitrary element in the structure of $E(Y_{ij})$, and it can be removed by introducing the restrictions

$$\Sigma\rho_i = \Sigma\kappa_j = 0.$$

Consider the hypothesis $\mathscr{H}_\rho$, according to which

$$\rho_1 = \rho_2 = \ldots = \rho_c.$$

The restriction $\Sigma\rho_i = 0$ implies that the common value of the row contributions is zero. Similarly, $\mathscr{H}_\kappa$ is the hypothesis that all the column contributions are zero. We proceed to devise tests of $\mathscr{H}_\rho$ and $\mathscr{H}_\kappa$.

The analysis is expressed in terms of averages taken over rows and columns. We write

$$\begin{aligned} y_{io} &= \Sigma_j y_{ij}, & y_{i\cdot} &= y_{io}/d, \\ y_{oj} &= \Sigma_i y_{ij}, & y_{\cdot j} &= y_{oj}/c, \\ y_{oo} &= \Sigma\Sigma y_{ij}, & y_{\cdot\cdot} &= y_{oo}/cd. \end{aligned}$$

This is a standard notation and will be used for other arrays without further comment. Take the identity

$$\begin{aligned}(y_{ij}-\mu-\rho_i-\kappa_j) \equiv (y_{\cdot\cdot}-\mu)+(y_{i\cdot}-y_{\cdot\cdot}-\rho_i)+(y_{\cdot j}-y_{\cdot\cdot}-\kappa_j) \\ +(y_{ij}-y_{i\cdot}-y_{\cdot j}+y_{\cdot\cdot}),\end{aligned}$$

square both sides and sum over i and j. All the product terms vanish. For example,

$$\begin{aligned}\Sigma\Sigma(y_{i\cdot}-y_{\cdot\cdot}-\rho_i)(y_{ij}-y_{i\cdot}-y_{\cdot j}+y_{\cdot\cdot}) \\ = \Sigma_i\{(y_{i\cdot}-y_{\cdot\cdot}-\rho_i)\Sigma_j(y_{ij}-y_{i\cdot}-y_{\cdot j}+y_{\cdot\cdot})\},\end{aligned}$$

and the inner sum is zero by definition of the averages involved. We thus obtain the basic identity

$$\begin{aligned}\Sigma\Sigma(y_{ij}-\mu-\rho_i-\kappa_j)^2 \equiv cd(y_{\cdot\cdot}-\mu)^2+d\Sigma(y_{i\cdot}-y_{\cdot\cdot}-\rho_i)^2 \\ +c\Sigma(y_{\cdot j}-y_{\cdot\cdot}-\kappa_j)^2+\Sigma\Sigma(y_{ij}-y_{i\cdot}-y_{\cdot j}+y_{\cdot\cdot})^2,\end{aligned}$$

and there is a parallel identity for the variates $\{Y_{ij}\}$.

According to the principle of least squares, the estimators of μ, $\{\rho_i\}$, and $\{\kappa_j\}$ are the values which minimize

$$W = \Sigma\Sigma(Y_{ij}-\mu-\rho_i-\kappa_j)^2.$$

In view of the identity just derived, the least squares estimators are

$$\hat{\mu} = Y.., \qquad \hat{\rho}_i = Y_{i}. - Y.., \qquad \hat{\kappa}_j = Y._j - Y..,$$

and the sum of squared residuals is

$$R = \Sigma\Sigma(Y_{ij} - Y_{i}. - Y._j + Y..)^2.$$

Under the hypothesis $\mathscr{H}_\rho$, according to which $\rho_i = 0$ for all i, the relative minimum of W is the absolute minimum of W_ρ, where

$$W_\rho = cd(Y.. - \mu)^2 + d\Sigma(Y_{i}. - Y..)^2 + c\Sigma(Y._j - Y.. - \kappa_j)^2 + R.$$

The minimum is attained when $\mu = \hat{\mu}$ and $\kappa_j = \hat{\kappa}_j$. Denote by Q_ρ the sum of squares for testing $\mathscr{H}_\rho$, and we find that

$$Q_\rho = d\Sigma(Y_{i}. - Y..)^2.$$

Similarly, the sum of squares for testing $\mathscr{H}_\kappa$ is

$$Q_\kappa = c\Sigma(Y._j - Y..)^2.$$

Let Q_μ be the sum of squares for testing that $\mu = 0$. Then

$$Q_\mu = cdY^2..,$$

known as the **correction factor** because of the way it enters into the calculations described below. By taking μ, $\{\rho_i\}$, and $\{\kappa_j\}$ all zero in the identity for W, we obtain

$$\Sigma\Sigma Y_{ij}^2 = Q_\mu + Q_\rho + Q_\kappa + R.$$

The last result implies the following computational routine. First, form the row totals and column totals, and check that the sum of row totals is equal to the sum of column totals, since both sums equal the grand total. Calculate $Q_\mu = Y_{00}^2/cd$. The identities

$$\Sigma Y_{i0}^2/d = Q_\mu + Q_\rho \quad \text{and} \quad \Sigma Y_{0j}^2/c = Q_\mu + Q_\kappa$$

give Q_ρ and Q_κ respectively. Then R is derived by subtraction of $Q_\mu + Q_\rho + Q_\kappa$ from $\Sigma\Sigma Y_{ij}^2$.

Finally, we turn to the relevant distribution theory. The assumptions about $\{Y_{ij}\}$ can be stated as

$$Y_{ij} = \mu + \rho_i + \kappa_j + \sigma Z_{ij} \quad (i = 1, 2, \ldots, c;\ j = 1, 2, \ldots, d),$$

where $\{Z_{ij}\}$ are mutually independent standard normal variates. Thus

$$Y_{i}. = \mu + \rho_i + \sigma Z_{i}., \quad Y._j = \mu + \kappa_j + \sigma Z._j,$$

and

$$Y_{..} = \mu + \sigma Z_{..}.$$

When these expressions are inserted in the basic identity the result is

$$W/\sigma^2 \equiv cdZ^2_{..} + d\Sigma(Z_{i.} - Z_{..})^2 + c\Sigma(Z_{.j} - Z_{..})^2$$
$$+\Sigma\Sigma(Z_{ij} - Z_{i.} - Z_{.j} + Z_{..})^2.$$

For a given value of i, consider an orthonormal linear transformation from $Z_{i1}, Z_{i2}, \ldots, Z_{id}$ to $U_{i1}, U_{i2}, \ldots, U_{id}$ such that $U_{i1} = d^{-\frac{1}{2}}Z_{i0}$. Make the same transformation for each i, and write $T_{1j} = c^{-\frac{1}{2}}U_{0j}$. Thus

$$T_{11} = c^{-\frac{1}{2}}U_{01} = (cd)^{-\frac{1}{2}}Z_{00},$$

and $T_{11}, T_{12}, \ldots, T_{1d}$ form an orthonormal linear transformation of $c^{-\frac{1}{2}}Z_{01}, c^{-\frac{1}{2}}Z_{02}, \ldots, c^{-\frac{1}{2}}Z_{0d}$. Make another orthonormal linear transformation from $U_{11}, U_{21}, \ldots, U_{c1}$ to $T_{11} = c^{-\frac{1}{2}}U_{01}$ and $T_{21}, \ldots, T_{c1}$. The transformation from $\{Z_{ij}\}$ to $T_{11}, T_{12}, \ldots, T_{1d}$, $T_{21}, \ldots, T_{c1}$ is orthonormal. We can therefore introduce further variates $\{T_{ij}\}$ for $i = 2, \ldots, c$ and $j = 2, \ldots, d$ such that the transformation from $\{Z_{ij}\}$ to $\{T_{ij}\}$ for all i and j is orthonormal. The two sets of variates are connected by

$$d^{-1}\Sigma Z^2_{i0} = \Sigma T^2_{i1}, \quad c^{-1}\Sigma Z^2_{0j} = \Sigma T^2_{1j}, \quad \text{and} \quad \Sigma\Sigma Z^2_{ij} = \Sigma\Sigma T^2_{ij}.$$

Consequently the terms on the right in the identity for W/σ^2 can be expressed as $cdZ^2_{..} = T^2_{11}$,

$$d\sum_{i=1}^{c}(Z_{i.} - Z_{..})^2 = \sum_{i=2}^{c} T^2_{i1}, \quad c\sum_{j=1}^{d}(Z_{.j} - Z_{..})^2 = \sum_{j=2}^{d} T^2_{1j},$$

and

$$\sum_{i=1}^{c}\sum_{j=1}^{d}(Z_{ij} - Z_{i.} - Z_{.j} + Z_{..})^2 = \sum_{i=2}^{c}\sum_{j=2}^{d} T^2_{ij}.$$

Since $\{T_{ij}\}$ are mutually independent standard normal variates, we conclude that these terms are mutually independent and distributed as $\chi^2_1, \chi^2_{c-1}, \chi^2_{d-1}$, and $\chi^2_{(c-1)(d-1)}$ respectively.

The sum of squares Q_ρ can be written as

$$Q_\rho = d\Sigma\{\rho_i + \sigma(Z_{i.} - Z_{..})\}^2.$$

This expression shows that Q_ρ/σ^2 is χ^2_{c-1} when $\mathscr{H}_\rho$ is true. Similar results hold for the other sums of squares, and the sum of squared

TABLE 14

Analysis of variance for two-way layout

Source of variation	*Sums of squares*	*Degrees of freedom*	*Mean squares*
Mean	Q_μ	1	Q_μ
Rows	Q_ρ	$c-1$	$Q_\rho/(c-1)$
Columns	Q_κ	$d-1$	$Q_\kappa/(d-1)$
Residual	R	$(c-1)(d-1)$	$R/(c-1)(d-1)$
Total	$\Sigma\Sigma Y_{ij}{}^2$	cd	

residuals R is such that R/σ^2 is $\chi^2_{(c-1)(d-1)}$. The complete analysis of variance is set out in Table 14. We test $\mathscr{H}_\rho$, for example, by referring the variance-ratio $(d-1)\,Q_\rho/R$ to a table of $F_{c-1,\,(c-1)(d-1);\,\varepsilon}$.

EXERCISES 9

1. $E(Y_i) = \alpha+\beta(x_i-\bar{x})$ and $\text{var}\,(Y_i) = \sigma^2$ $(i = 1, 2, \ldots, n)$, where $Y_1, Y_2, \ldots, Y_n$ are mutually independent. Show that the least squares estimator $\hat{\alpha}$ is the best linear unbiased estimator of α.

2. $E(Y_i) = \alpha+\beta(x_i-\bar{x})$ and $\text{var}\,(Y_i) = \sigma^2$ $(i = 1, 2, \ldots, n)$, where $Y_1, Y_2, \ldots, Y_n$ are mutually independent. Let n be an even number, say $2h$, and write $D_j = Y_{n-j+1}-Y_j$. Then

$$E(D_j) = \beta(x_{n-j+1}-x_j) \text{ and } \text{var}\,(D_j) = 2\sigma^2 \; (j = 1, 2, \ldots, h),$$

where $D_1, D_2, \ldots, D_h$ are mutually independent. Denote by β^* the least squares estimator of β in terms of $\{D_j\}$. Show that

$$\text{var}\,(\beta^*) = \sigma^2\{\Sigma(x_i-\bar{x})^2-\tfrac{1}{2}\Sigma(x_{n-j+1}+x_j-2\bar{x})^2\}^{-1}.$$

Deduce that $\text{var}\,(\beta^*) \geqslant \text{var}\,(\hat{\beta})$ with equality if and only if $x_{n-j+1}+x_j$ is independent of j.

3. $E(Y_i) = \alpha+\beta(x_i-\bar{x})$ and $\text{var}\,(Y_i) = \sigma^2$ $(i = 1, 2, \ldots, n)$. Each pair of variates has correlation coefficient ρ. Show that

$$E(R) = (n-2)\sigma^2(1-\rho).$$

4. Regression lines arise from k independent experiments. The statistical model is as follows.

$$E(Y_{ij}) = \alpha_j+\beta_j(x_{ij}-\bar{x}_j) \quad\text{and}\quad \text{var}\,(Y_{ij}) = \sigma^2,$$

for $i = 1, 2, \ldots, n_j$ and $j = 1, 2, \ldots, k$, where $\bar{x}_j = \Sigma_i x_{ij}/n_j$. Write

$f_j = \Sigma_i(x_{ij}-\bar{x}_j)^2$, $G_j = \Sigma_i Y^2_{ij}-(\Sigma_i Y_{ij})^2/n_j$, $H_j = \Sigma_i(x_{ij}-\bar{x}_j)\,Y_{ij}$,

and

$$W = \Sigma\Sigma\{Y_{ij}-\alpha_j-\beta_j(x_{ij}-\bar{x}_j)\}^2.$$

Show that the sum of squared residuals, and the sum of squares for testing $\mathscr{H}: \beta_1 = \beta_2 = \ldots = \beta_k$, are given respectively by

$$R = \Sigma(G_j-H_j^2/f_j) \quad \text{and} \quad Q = \Sigma(H_j^2/f_j)-(\Sigma H_j)^2/\Sigma f_j.$$

5. The conditions are the same as in the previous exercise. In addition, the variates $\{Y_{ij}\}$ are normally and independently distributed. Show that R/σ^2 is χ^2_{n-2k}, where $n = \Sigma n_j$, and that Q/σ^2 is independently χ^2_{k-1} when $\mathscr{H}$ is true. Prove that the variance-ratio test of $\mathscr{H}$ reduces when $k = 2$ to the test given in §**45**.

6. The conditions are the same as in the previous exercise. Suppose that $\mathscr{H}$ is accepted. Find the sum of squares for testing that

$$\alpha_j = \alpha+\beta\bar{x}_j \quad (j = 1, 2, \ldots, k),$$

where the values of α and β are unspecified.

7. The variates $Y_1, Y_2, \ldots, Y_n$ are mutually independent and such that var $(Y_i) = \sigma^2/d_i$, where $d_1, d_2, \ldots, d_n$ are known. Suppose that

$$E(Y_i) = \alpha+\beta(x_i-\bar{x}) \quad (i = 1, 2, \ldots, n),$$

where $\bar{x} = \Sigma d_i x_i/\Sigma d_i$. Define

$$U(\alpha, \beta) = \Sigma d_i\{Y_i-\alpha-\beta(x_i-\bar{x})\}^2.$$

Let $\hat{\alpha}$ and $\hat{\beta}$ now be the values of α and β respectively at which the absolute minimum of $U(\alpha, \beta)$ occurs. Show that $\hat{\alpha}$ and $\hat{\beta}$ are the best linear unbiased estimators of α and β respectively. Prove also that

$$E\{U(\hat{\alpha}, \hat{\beta})\} = (n-2)\sigma^2.$$

8. For the data of Table 13 (p. 157), let ρ_i be the contribution of the ith rat and κ_j that of the jth meal. Calculate the analysis of variance and test the hypotheses $\mathscr{H}_\rho$ and $\mathscr{H}_\kappa$. Comment on the results.

9. Under the conditions of §**46**, show that the expectations of the sums of squares in Table 14 are respectively $cd\mu^2+\sigma^2$, $d\Sigma\rho_i^2+(c-1)\sigma^2$, $c\Sigma\kappa_j^2+(d-1)\sigma^2$, and $(c-1)(d-1)\sigma^2$.

10. A random effects model for a two-way layout arises when the row and column contributions are all random. Suppose that

$$Y_{ij} = \mu+\rho_i+\kappa_j+\sigma Z_{ij} \quad (i = 1, 2, \ldots, c;\ j = 1, 2, \ldots, d),$$

where $\{\rho_i\}$, $\{\kappa_j\}$, and $\{Z_{ij}\}$ are mutually independent normal variates with zero means and such that

$$\text{var}(\rho_i) = \sigma_\rho^2, \ \text{var}(\kappa_j) = \sigma_\kappa^2, \ \text{var}(Z_{ij}) = 1.$$

Show that Q_ρ, Q_κ, and R are mutually independent and distributed as follows: Q_ρ is $(\sigma^2+d\sigma_\rho^2)\chi_{c-1}^2$, Q_κ is $(\sigma^2+c\sigma_\kappa^2)\chi_{d-1}^2$, and R is $\sigma^2\chi^2_{(c-1)(d-1)}$. Derive a test of the hypothesis $\sigma_\rho^2 = 0$ and a confidence interval for σ_κ^2/σ^2.

CHAPTER 10
ASYMPTOTIC DISTRIBUTIONS

§47. Convergence of variates

Suppose that a random sample is taken from the distribution of a continuous variate. The exact distribution of a sample statistic can be determined only if the relevant multiple integral can be evaluated, or if the transform of the characteristic function can be found. Similar conditions apply to discrete variates. The operations called for can sometimes be successfully completed, as for example in the problems of chapter 8, but in general both methods of approach lead to difficulties. However, the exact statements which elude us in finite samples can usually be obtained in relation to the asymptotic properties of a statistic as the sample size tends to infinity. We are then concerned with a sequence of variates indexed by the sample size, and the question naturally arises as to how the convergence of this sequence may be defined. Several definitions are available, and the conditions under which they obtain are discussed in what follows.

We say that a sequence of variates $T_1, T_2, \ldots$ **converges in mean square** to a constant θ if

$$\lim_{n \to \infty} E(T_n - \theta)^2 = 0.$$

This is the simplest type of convergence to establish in practice.

Example 1. $X_1, X_2, \ldots$ is a sequence of mutually independent variates, each with mean μ and variance σ^2, both finite. Define $M_n = \Sigma_1^n X_i/n$. Then $E(M_n) = \mu$ and $E(M_n - \mu)^2 = \sigma^2/n$. Thus M_n converges in mean square to μ. As a particular application, consider a sequence of independent experiments, in each of which the probability is θ that an event $\mathscr{E}$ occurs. Define $X_n = 1$ if $\mathscr{E}$ occurs in the

nth experiment and $X_n = 0$ if $\mathscr{E}$ does not occur in the nth experiment, so that M_n is the frequency-ratio for $\mathscr{E}$. Here $\mu = \theta$ and $\sigma^2 = \theta(1-\theta)$. Thus the frequency-ratio for $\mathscr{E}$ converges in mean square to θ, which gives a more precise statement of the assertion in §2 that a frequency-ratio tends to a probability in the limit.

We say that a sequence $T_1, T_2, \ldots$ **converges in probability** to θ if

$$\lim_{n \to \infty} p(|T_n - \theta| > \delta) = 0,$$

for any $\delta > 0$. An equivalent condition is

$$\lim_{n \to \infty} p(|T_n - \theta| \leqslant \delta) = 1.$$

We shall prove that convergence in mean square implies convergence in probability. Suppose that $(T_n - \theta)$ is a continuous variate with density function $g_n(y)$. The proof when $(T_n - \theta)$ has a discrete distribution is a straightforward analogy. By definition,

$$\begin{aligned} E(T_n - \theta)^2 &= \int_{-\infty}^{\infty} y^2 g_n(y)\,dy, \\ &= \int_{|y| \leqslant \delta} y^2 g_n(y)\,dy + \int_{|y| > \delta} y^2 g_n(y)\,dy. \end{aligned}$$

Each term on the right is non-negative. Therefore

$$E(T_n - \theta)^2 \geqslant \int_{|y| > \delta} y^2 g_n(y)\,dy \geqslant \delta^2 \int_{|y| > \delta} g_n(y)\,dy.$$

The final integral is $p(|T_n - \theta| > \delta)$, so that

$$p(|T_n - \theta| > \delta) \leqslant E(T_n - \theta)^2/\delta^2.$$

This is known as Chebyshev's inequality. If T_n converges in mean square to θ, then $E(T_n - \theta)^2 \to 0$ as $n \to \infty$, and hence also $E(T_n - \theta)^2/\delta^2 \to 0$ for any fixed δ, however small. Thus $p(|T_n - \theta| > \delta)$ is bounded above by a quantity which tends to zero as $n \to \infty$. We conclude that T_n converges in probability to θ.

Suppose that $T_1, T_2, \ldots$ is a sequence which converges in probability to θ and let $h(x)$ be a continuous function. According to Slutsky's theorem, $h(T_n)$ converges in probability to $h(\theta)$. The proof is as follows. Given $\varepsilon > 0$, we can find δ such that $|h(T_n) - h(\theta)| \leqslant \varepsilon$ whenever $|T_n - \theta| \leqslant \delta$. For any $\delta > 0$,

$$\lim_{n \to \infty} p(|T_n - \theta| \leqslant \delta) = 1.$$

Hence also

$$\lim_{n \to \infty} p\{|\, h(T_n) - h(\theta)\,| \leqslant \varepsilon\} = 1,$$

which shows that $h(T_n)$ converges in probability to $h(\theta)$.

Example 2. $X_1, X_2, \ldots$ is a sequence of variates which are mutually independent and identically distributed. Suppose that the common fourth central moment μ_4 is finite, in which case the mean μ and variance σ^2 are also finite. Define

$$S_n^2 = \sum_{i=1}^{n} (X_i - M_n)^2/(n-1).$$

In §**28** we showed that $E(S_n^2) = \sigma^2$, and noted that

$$\text{var}\,(S_n^2) = (\mu_4 - \mu_2^2)/n + 2\mu_2^2/n(n-1).$$

Thus S_n^2 converges in probability to σ^2. Taking positive square roots, we find that S_n converges in probability to σ.

The possibility arises that the limit of a sequence of variates $T_1, T_2, \ldots$ is another variate T. Denote by $F_n(t)$ and $F(t)$ the distribution functions of T_n and T respectively. We say that T_n **converges in distribution** to T if

$$\lim_{n \to \infty} F_n(t) = F(t)$$

at every point where F is continuous. Convergence in distribution can be established with the assistance of the following results.

Suppose that $h(x)$ is a function with a continuous positive derivative, so that the equation $h(x) = t$ has a unique solution $x = k(t)$, and the inequality $h(x) \leqslant t$ is equivalent to $x \leqslant k(t)$. The distribution function of $h(T_n)$ is

$$p\{h(T_n) \leqslant t\} = p\{T_n \leqslant k(t)\} = F_n\{k(t)\},$$

and the distribution function of $h(T)$ is similarly $F\{k(t)\}$. Hence $h(T_n)$ converges in distribution to $h(T)$. The conditions on h can be relaxed as in §**21**.

Denote the characteristic function of T_n by $D_n(q)$. According to the Lévy-Cramér theorem, a necessary and sufficient condition that T_n converges in distribution to T is that $D_n(q)$ converges to a limit $D(q)$ which is continuous at $q = 0$, and when the condition holds $D(q)$ is the characteristic function of T. A proof is beyond the scope of this book. The theorem enables us to infer convergence in distribution from the convergence of a sequence of characteristic functions. Since there are many problems where the characteristic function is more easily determined than the distribution function, we thus have a valuable method of deriving the limit of a sequence of distribution functions without calculating any member of the sequence.

Let $\{A_n\}$ be a sequence which converges in distribution to A, and let $\{B_n\}$ be a sequence which converges in probability to a positive constant b. Define $T_n = A_n/B_n$. According to Cramér's theorem, the sequence $\{T_n\}$ converges in distribution to A/b. A proof is as follows.

$$p(T_n \leqslant t) = p(T_n \leqslant t, \mid B_n - b \mid \leqslant \delta) + p(T_n \leqslant t, \mid B_n - b \mid > \delta).$$

Under the conditions $A_n/B_n \leqslant t$ and $\mid B_n - b \mid \leqslant \delta$, the maximum value of A_n is $t(b+\delta)$. Hence

$$p(T_n \leqslant t, \mid B_n - b \mid \leqslant \delta) \leqslant p\{A_n \leqslant t(b+\delta), \mid B_n - b \mid \leqslant \delta\}.$$

Make the substitution

$$p\{A_n \leqslant t(b+\delta), \mid B_n - b \mid \leqslant \delta) \\ = p\{A_n \leqslant t(b+\delta)\} - p\{A_n \leqslant t(b+\delta), \mid B_n - b \mid > \delta\},$$

and we obtain

$$p(T_n \leqslant t) \leqslant p\{A_n \leqslant t(b+\delta)\} + p\{T_n \leqslant t, \mid B_n - b \mid > \delta\} \\ - p\{A_n \leqslant t(b+\delta), \mid B_n - b \mid > \delta\}.$$

Fix δ at a value which is arbitrarily small. The total of the second and third terms on the right is at most equal to $p(\mid B_n - b \mid > \delta)$, and therefore tends to zero as $n \to \infty$. If the distribution function of A is continuous at tb, then

$$\lim_{n \to \infty} p(T_n \leqslant t) \leqslant p\{A \leqslant t(b+\delta)\}.$$

By the same method of argument,

$$\lim_{n \to \infty} p(T_n \leqslant t) \geqslant p\{A \leqslant t(b-\delta)\}.$$

Since δ is arbitrarily small, the conclusion is that T_n converges in distribution to A/b.

We say that the vector sequence $(A_1, B_1, \ldots, C_1)$, $(A_2, B_2, \ldots, C_2)$, ... converges in distribution to the vector $(A, B, \ldots, C)$ if the joint distribution function of $A_n, B_n, \ldots, C_n$ converges to the joint distribution function of $A, B, \ldots, C$ at every continuity point. Suppose that $h(x, y, \ldots, z)$ is a continuous function. According to

Sverdrup's theorem, $h(A_n, B_n, \ldots, C_n)$ converges in distribution to $h(A, B, \ldots, C)$. The proof is omitted.

§48. Central limit theorem

Suppose that X is a variate with mean μ and variance σ^2, both finite. Let $X_1, X_2, \ldots$ be a sequence of mutually independent variates, each distributed like X. Denote $\Sigma_1^n X_i/n$ by M_n, and define

$$Z_n = n^{\frac{1}{2}}(M_n - \mu)/\sigma.$$

Then Z_n has mean 0 and variance 1 for all n. According to the central limit theorem, Z_n converges in distribution to $\mathcal{N}(0, 1)$. Thus

$$\lim_{n\to\infty} F_n(z) = \Phi(z)$$

for all z, where F_n is the distribution function of Z_n, and Φ is that of $\mathcal{N}(0, 1)$. We have already seen that Z_n is $\mathcal{N}(0, 1)$ for all n when X is normally distributed. The central limit theorem states that Z_n is asymptotically $\mathcal{N}(0, 1)$ whatever the distribution of X.

We shall sketch the proof. Write $Z = (X-\mu)/\sigma$. Denote the moment generating function of Z by $J(\omega)$, and suppose that $J(\omega)$ exists throughout an interval containing $\omega = 0$. Since Z has mean 0 and variance 1,

$$J(\omega) = 1 + \tfrac{1}{2}\omega^2 + \rho(\omega^2),$$

where $\rho(x)$ is a quantity such that $\rho(x)/x \to 0$ as $x \to 0$. Now Z_n can be expressed in the form $\Sigma(X_i - \mu)/\sigma n^{\frac{1}{2}}$ and hence its moment generating function is

$$\{J(\omega/n^{\frac{1}{2}})\}^n = \{1 + \omega^2/2n + \rho(\omega^2/n)\}^n.$$

Fix ω and let $n \to \infty$. From the definition of ρ, $\rho(\omega^2/n)$ is such that $n\rho(\omega^2/n)/\omega^2 \to 0$, and hence also $n\rho(\omega^2/n) \to 0$. Thus

$$\lim_{n\to\infty} \{J(\omega/n^{\frac{1}{2}})\}^n = \exp(\tfrac{1}{2}\omega^2)$$

for all ω. This argument remains valid when J is replaced by the characteristic function of Z, without any restriction on Z other than $E(Z) = 0$ and var $(Z) = 1$. The equivalence between a sequence of distribution functions and the corresponding sequence of characteristic functions now implies that Z_n converges in distribution to $\mathcal{N}(0, 1)$.

The practical importance of the central limit theorem resides in the implication that Z_n is approximately $\mathcal{N}(0, 1)$ when n is large. Equivalent statements can be made in terms of M_n or $T_n = \Sigma_1^n X_i$. Thus M_n is approximately $\mathcal{N}(\mu, \sigma^2/n)$ and T_n is approximately $\mathcal{N}(n\mu, n\sigma^2)$.

Example. X is distributed as χ_1^2. Since $E(X) = 1$ and var $(X) = 2$, we have $E(T_n) = n$ and var $(T_n) = 2n$. A reference to §**23** shows that T_n is distributed as χ_n^2. Hence the values of $p(T_n \leqslant t)$ can be obtained exactly from tables of χ_n^2, or approximately from

$$p(T_n \leqslant t) = \Phi\{(t-n)/(2n)^{\frac{1}{2}}\}.$$

The exact and approximate results are compared below for $n = 60$ and various values of t.

	t	45	50	55	60	65	70	75
Exact	$p(T_{60} \leqslant t)$	0·075	0·182	0·341	0·524	0·693	0·823	0·908
Approx.	$p(T_{60} \leqslant t)$	0·085	0·181	0·324	0·500	0·676	0·819	0·914

We conclude with two illustrations of the consequences of the central limit theorem for statistical methodology. First, suppose that a sample of size n is taken from a distribution with finite μ_4. The mean μ and variance σ^2 are both unknown. Denote the sample mean and variance by M and S^2 respectively. When $n \to \infty$, $n^{\frac{1}{2}}(M-\mu)/\sigma$ converges in distribution to $\mathcal{N}(0, 1)$, and S converges in probability to σ. The variate

$$U_n = n^{\frac{1}{2}}(M-\mu)/S$$

is the ratio of $n^{\frac{1}{2}}(M-\mu)/\sigma$ to S/σ. Since S/σ converges in probability to 1, an application of Cramér's theorem shows that U_n converges in distribution to $\mathcal{N}(0, 1)$. In large samples, U_n is approximately $\mathcal{N}(0, 1)$, and by the standard argument $m \pm u_{\frac{1}{2}\varepsilon} s/n^{\frac{1}{2}}$ is a confidence interval for μ with confidence coefficient approximately $1-\varepsilon$. This result is independent of assumptions about the mathematical form of the distribution of X such as we made in §**38**.

Secondly, suppose that two samples are available, where the ith sample has size n_i and is taken from a distribution with mean μ_i and variance σ_i^2 $(i = 1, 2)$. We assume that the fourth central moments are both finite, and denote the sample means by M_1, M_2 and the sample variances by S_1^2, S_2^2. When $n_1, n_2 \to \infty$, a straightforward extension of the central limit theorem shows that

$$\{(M_1-M_2)-(\mu_1-\mu_2)\}(\sigma_1^2/n_1+\sigma_2^2/n_2)^{-\frac{1}{2}}$$

converges in distribution to $\mathcal{N}(0, 1)$. The variate

$$(S_1^2/n_1+S_2^2/n_2) \,/\, (\sigma_1^2/n_1+\sigma_2^2/n_2)$$

converges in probability to 1 (see exercise 1). Write

$$U_{n_1, n_2} = \{(M_1-M_2)-(\mu_1-\mu_2)\}(S_1^2/n_1+S_2^2/n_2)^{-\frac{1}{2}}.$$

Then U_{n_1,n_2} converges in distribution to $\mathcal{N}(0, 1)$ as $n_1, n_2 \to \infty$. This result is independent of the assumption that $\sigma_1^2 = \sigma_2^2$, so that confidence intervals and significance tests based on U_{n_1,n_2} are preferable in large samples to the procedures developed for two normal samples in §39.

§49. Limit theory for discrete distributions

The central limit theorem applies to any variate T_n which can be expressed in the form $\Sigma_1^n X_i$, where $X_1, X_2, \ldots$ are independently and identically distributed with finite mean and variance. For example, suppose that the possible outcomes of a trial are success, with probability θ, and failure, with probability $1-\theta$. Let X be the number of successes in a trial, so that

$$p(X = 1) = \theta \quad \text{and} \quad p(X = 0) = 1-\theta.$$

The mean and variance of X are θ and $\theta(1-\theta)$ respectively. Here T_n is the number of successes in an experiment which consists of n independent trials. Consequently T_n has the binomial distribution

$$p(T_n = j) = \binom{n}{j}\theta^j(1-\theta)^{n-j} \quad (j = 0, 1, \ldots, n).$$

Since $E(T_n) = n\theta$ and var $(T_n) = n\theta(1-\theta)$, the central limit theorem implies that

$$p(T_n \leqslant j) \simeq \Phi[(j-n\theta)/\{n\theta(1-\theta)\}^{\frac{1}{2}}].$$

However, the asymptotic properties of binomial, Poisson, and hypergeometric distributions can be obtained directly by a detailed examination of the individual probabilities. An essential tool is Stirling's formula

$$k! \sim (2\pi)^{\frac{1}{2}}k^{k+\frac{1}{2}}e^{-k}.$$

Consider a variate R_n with a hypergeometric distribution given by

$$p(R_n = r_1) = \binom{n_1}{r_1}\binom{n_2}{r_2}\Big/\binom{n}{r},$$

where $n_1+n_2 = n$ and $r_1+r_2 = r$. The distribution was introduced in §**8** and our previous notation has been modified for convenience. Let n_1, n_2, r and $n-r$ all tend to infinity in such a way that $n_1/n \to \psi$ and $r/n \to \theta$. Then

$$Z_n = (R_n - n\psi\theta)/\{n\psi(1-\psi)\theta(1-\theta)\}^{\frac{1}{2}}$$

converges in distribution to $\mathcal{N}(0, 1)$ as $n \to \infty$. This result cannot be proved by a straightforward appeal to the central limit theorem because R_n is not a sum of mutually independent variates. A proof based on Stirling's formula is as follows. Denote $p(R_n = r_1)$ by p and replace each factorial in

$$p = n_1!\, n_2!\, r!\, (n-r)! \,/\, n!\, r_1!\, (n_1-r_1)!\, r_2!\, (n_2-r_2)!$$

by the corresponding value of $(2\pi)^{\frac{1}{2}} k^{k+\frac{1}{2}} \mathrm{e}^{-k}$. There are four contributions from $(2\pi)^{\frac{1}{2}}$ in the numerator of p and five in the denominator, so that this factor of $k!$ gives $(2\pi)^{-\frac{1}{2}}$ in p. The contributions from e^{-k} cancel out because

$$n_1+n_2+r+(n-r) = n+r_1+(n_1-r_1)+r_2+(n_2-r_2).$$

Consider next the contributions to p from $k^{k+\frac{1}{2}}$. They are more conveniently treated as contributions to $\log p$ from $(k+\frac{1}{2}) \log k$. Substitute $n_1 r/n + x$ for r_1, whence

$$(r_1+\tfrac{1}{2}) \log r_1 = (r_1+\tfrac{1}{2}) \log (n_1 r/n) + (n_1 r/n + x + \tfrac{1}{2}) \log (1 + xn/n_1 r).$$

We expand the second term on the right assuming that x is $0(n^{\frac{1}{2}})$, and subsequently show that the probability for other values of x is negligible. Thus

$$(r_1+\tfrac{1}{2}) \log r_1 = (r_1+\tfrac{1}{2})(\log n_1 + \log r - \log n) + x + nx^2/2n_1 r + 0(n^{-\frac{1}{2}})$$

The definition of x implies that $r_2 = n_2 r/n - x$,

$$n_1 - r_1 = n_1(n-r)/n - x \quad \text{and} \quad n_2 - r_2 = n_2(n-r)/n + x.$$

Expansions of $(k+\frac{1}{2}) \log k$ similar to that when $k = r_1$ are therefore obtainable when $k = r_2$, $n_1 - r_1$ and $n_2 - r_2$. We now collect the contributions to $\log p$ from $(k+\frac{1}{2}) \log k$. The coefficient of $\log n_1$ is

$$(n_1+\tfrac{1}{2}) - (r_1+\tfrac{1}{2}) - (n_1 - r_1 + \tfrac{1}{2}) = -\tfrac{1}{2}.$$

Those of $\log n_2$, $\log r$ and $\log (n-r)$ are also $-\frac{1}{2}$, while that of $\log n$ is

$$-(n+\tfrac{1}{2}) + (r_1+\tfrac{1}{2}) + (n_1 - r_1 + \tfrac{1}{2}) + (r_2+\tfrac{1}{2}) + (n_2 - r_2 + \tfrac{1}{2}) = 3/2.$$

The coefficient of x is zero, and that of x^2 is

$$\frac{n}{2}\left\{\frac{1}{n_1 r}+\frac{1}{n_2 r}+\frac{1}{n_1(n-r)}+\frac{1}{n_2(n-r)}\right\}=\frac{n^3}{2n_1n_2r(n-r)}.$$

When all the terms in $\log p$ are collected, the result is

$$\log p = -\tfrac{1}{2}\log\{2\pi n_1 n_2 r(n-r)/n^3\} - x^2 n^3/2n_1n_2r(n-r) + 0(n^{-\frac{1}{2}}).$$

We finally take the limit as n_1, n_2, r and $n-r$ tend to infinity under the conditions stated, and obtain

$$p(R_n = r_1)$$
$$\sim \{2\pi n\psi(1-\psi)\theta(1-\theta)\}^{-\frac{1}{2}} \exp\{-(r_1 - n\psi\theta)^2/2n\psi(1-\psi)\theta(1-\theta)\}.$$

This shows that Z_n is asymptotically $\mathcal{N}(0, 1)$, and confirms that values of x which have an order larger than $0(n^{\frac{1}{2}})$ can be neglected.

Similar results can be proved for a binomial distribution with index n and parameter θ, and a Poisson distribution with parameter $n\lambda$. In all these cases, we find that the asymptotic distribution of an integer-valued variate V_n is given by

$$p(V_n = j) \sim (2\pi n\sigma^2)^{-\frac{1}{2}} \exp\{-(j-n\mu)^2/2n\sigma^2\}$$

for appropriate values of μ and σ^2. Thus individual probabilities are asymptotically equal to the ordinates of $\mathcal{N}(n\mu, n\sigma^2)$. Write

$$f(v) = (2\pi n\sigma^2)^{-\frac{1}{2}} \exp\{-(v-n\mu)^2/2n\sigma^2\}.$$

This is the density function of $\mathcal{N}(n\mu, n\sigma^2)$. For large n, $f(v)$ is almost linear over any interval of unit length, so that

$$\int_{k-\frac{1}{2}}^{k+\frac{1}{2}} f(v)\,dv \simeq f(k).$$

Since $p(V_n = k) \sim f(k)$, the distribution function of V_n is given by

$$p(V_n \leqslant j) \simeq \sum_{k=0}^{j} \int_{k-\frac{1}{2}}^{k+\frac{1}{2}} f(v)\,dv = \int_{-\frac{1}{2}}^{j+\frac{1}{2}} f(v)\,dv.$$

The last integral differs by a negligible amount from

$$\int_{-\infty}^{j+\frac{1}{2}} f(v)\,dv = \Phi\{(j+\tfrac{1}{2}-n\mu)/n^{\frac{1}{2}}\sigma\}.$$

We therefore obtain

$$p(V_n \leqslant j) \simeq \Phi\{(j+\tfrac{1}{2}-n\mu)/n^{\frac{1}{2}}\sigma\}.$$

In the same way

$$p(V_n \geqslant j) \simeq 1-\Phi\{(j-\tfrac{1}{2}-n\mu)/n^{\frac{1}{2}}\sigma\},$$

with the alternative representation

$$p(V_n \geqslant j) \simeq \Phi\{n\mu-j+\tfrac{1}{2})/n^{\frac{1}{2}}\sigma\}.$$

The $\pm\frac{1}{2}$ which appears in these expressions is known as a **correction for continuity.** Caution is necessary in the use of such approximations because the correction adjusts $\Phi\{(j-n\mu)/n^{\frac{1}{2}}\sigma\}$ by an amount which has the same order of magnitude, namely $0(n^{-\frac{1}{2}})$, as the error incurred when $p(V_n = k)$ is replaced by $f(k)$ for $k = 0, 1, \ldots, j$.

Example. When T_n has a binomial distribution with index n and parameter θ,

$$p(T_n \leqslant j) \simeq \Phi[(j+\tfrac{1}{2}-n\theta)/\{n\theta(1-\theta)\}^{\frac{1}{2}}].$$

The exact and approximate values of $p(T_n \leqslant j)$ are compared below for $n = 60$, $\theta = 0{\cdot}2$, and various values of j.

j	4	7	10	13	16	19
Exact $p(T_{60} \leqslant j)$	0·004	0·067	0·323	0·694	0·923	0·989
Approx. $p(T_{60} \leqslant j)$	0·008	0·073	0·314	0·686	0·927	0·992

§50. Limit theory for order statistics

Suppose that $X_1, X_2, \ldots, X_n$ are mutually independent variates each with density function $f(x)$ and distribution function $F(x)$. Arrange them in order of magnitude and denote the ordered variates by $X_{1n}, X_{2n}, \ldots, X_{nn}$, where

$$X_{1n} < X_{2n} < \ldots < X_{nn}.$$

Let ψ be a probability such that the equation $F(\xi) = \psi$ has a unique solution ξ. We shall find the asymptotic distribution of X_{kn} when k and n each tend to infinity in such a way that k is always the least integer which satisfies $k \geqslant n\psi$. According to §**30**, the distribution function of X_{kn} is given by

$$p(X_{kn} \leqslant y) = \sum_{i=k}^{n} \binom{n}{i} \{F(y)\}^i\{1-F(y)\}^{n-i}.$$

The terms on the right are probabilities of a binomial distribution with index n and parameter $F(y)$. An application of the central limit theorem shows that

$$p(X_{kn} \leqslant y) \simeq \Phi\left\{\frac{nF(y)-k}{[nF(y)\{1-F(y)\}]^{\frac{1}{2}}}\right\}.$$

Since k/n differs from $\psi = F(\xi)$ by less than $1/n$,

$$p(X_{kn} \leqslant y) \simeq \Phi\left\{\frac{F(y)-F(\xi)}{[F(y)\{1-F(y)\}/n]^{\frac{1}{2}}}\right\}.$$

Virtually the whole of a standard normal distribution lies between ± 5, say. Hence the distribution of X_{kn} ranges over values y such that $F(y)$ differs from $F(\xi)$ by an amount which is $0(n^{-\frac{1}{2}})$ with a probability close to 1. The same probability applies to all the following deductions from this premise. Because of the continuity of F, y differs from ξ by terms which are $0(n^{-\frac{1}{2}})$. Write $y = \xi + tn^{-\frac{1}{2}}$. Expanding in Taylor series, we have

$$F(y) = F(\xi)+tn^{-\frac{1}{2}}f(\xi)+0(n^{-1}),$$

and

$$F(y)\{1-F(y)\} = \psi(1-\psi)+0(n^{-\frac{1}{2}}).$$

The asymptotic form of the distribution of X_{kn} becomes

$$p(X_{kn} \leqslant y) \simeq \Phi[tf(\xi)/\{\psi(1-\psi)\}^{\frac{1}{2}}].$$

Denote $tf(\xi)/\{\psi(1-\psi)\}^{\frac{1}{2}}$ by u, whence

$$y = \xi+u\{\psi(1-\psi)/n\}^{\frac{1}{2}}/f(\xi).$$

We finally obtain

$$p[(X_{kn}-\xi)f(\xi)/\{\psi(1-\psi)/n\}^{\frac{1}{2}} \leqslant u] \simeq \Phi(u),$$

which shows that $(X_{kn}-\xi)f(\xi)/\{\psi(1-\psi)/n\}^{\frac{1}{2}}$ converges in distribution to $\mathcal{N}(0, 1)$ as $n \to \infty$.

When $\psi = \frac{1}{2}$, ξ is the population median. Thus the sequence of variates

$$X_{12}, X_{23}, X_{24}, X_{35}, X_{36}\ X_{47}, \ldots$$

is such that $2(X_{kn}-\xi)n^{\frac{1}{2}}f(\xi)$ converges in distribution to $\mathcal{N}(0, 1)$. The sequence of sample medians

$$\tfrac{1}{2}(X_{11}+X_{12}), X_{23}, \tfrac{1}{2}(X_{24}+X_{34}), X_{35}, \tfrac{1}{2}(X_{36}+X_{46}), \ldots$$

has the same limit property. Hence the sample median Q_n is asymptotically normal with mean ξ and variance $\{4nf^2(\xi)\}^{-1}$.

Example 1. A sample of size n is taken from $\mathcal{N}(\mu, \sigma^2)$, for which

$$f(x) = (2\pi\sigma^2)^{-\frac{1}{2}} \exp\{-(x-\mu)^2/2\sigma^2\}.$$

The median of this distribution is μ, and $f(\mu) = (2\pi\sigma^2)^{-\frac{1}{2}}$. Consequently Q_n is asymptotically $\mathcal{N}(\mu, \pi\sigma^2/2n)$. In large samples from a normal distribution with known variance, a confidence interval for μ with confidence coefficient approximately $1-\varepsilon$ is given by $q_n \pm u_{\frac{1}{2}\varepsilon}\sigma(\pi/2n)^{\frac{1}{2}}$. This is the shortest interval based on the sample median, and its length is $2u_{\frac{1}{2}\varepsilon}\sigma(\pi/2n)^{\frac{1}{2}}$. For the same confidence coefficient the shortest interval based on the sample mean has length $2u_{\frac{1}{2}\varepsilon}\sigma/n^{\frac{1}{2}}$. Thus the interval based on the median is longer by a factor of $(\pi/2)^{\frac{1}{2}} = 1{\cdot}25\ldots$, and for this reason the interval based on the mean is preferable in large samples. However, our approximation to the distribution of Q_n does not permit a comparison of the mean and median in samples which are small or moderate in size, and a further study of the distribution would be necessary.

Example 2. A sample of size n is taken from a double exponential distribution, where

$$f(x) = (2\lambda)^{-1} \exp\{-|(x-\mu)/\lambda|\} \quad (-\infty < x < \infty).$$

The median of this distribution is μ and $f(\mu) = (2\lambda)^{-1}$, so that Q_n is asymptotically $\mathcal{N}(\mu, \lambda^2/n)$. Since $E(X) = \mu$ and $\operatorname{var}(X) = 2\lambda^2$, M_n is asymptotically $\mathcal{N}(\mu, 2\lambda^2/n)$. In large samples from a distribution with known λ, the median is therefore clearly preferable to the mean as a statistic on which to base confidence intervals for μ.

§51. Limit theory for maximum likelihood estimators

The definition of likelihood in §32 can be applied whatever the joint distribution of the variates observed. Suppose that the joint distribution of $X_1, X_2, \ldots, X_n$ depends on a single parameter θ. Denote the likelihood of the observed values by l and let the corresponding variate be L. For example, if $X_1, X_2, \ldots, X_n$ are independently and identically distributed, then

$$L = f(X_1; \theta) f(X_2; \theta) \ldots f(X_n; \theta),$$

where $f(x; \theta)$ is the density function or probability distribution as the case may be. The maximum likelihood estimator $\hat{\theta}$ is obtained by solving the equation

$$\partial \log L/\partial\theta = 0.$$

According to the limit theorem for maximum likelihood estimators this equation has a solution $\hat{\theta}$ such that

$$(\hat{\theta}-\theta)\{-E(\partial^2 \log L/\partial\theta^2)\}^{\frac{1}{2}}$$

converges in distribution to $\mathcal{N}(0, 1)$, provided that certain conditions on the joint distribution of $X_1, X_2, \ldots, X_n$ are satisfied. We omit a statement of the conditions and a proof of the theorem. In large samples, $\hat{\theta}$ is approximately normal with mean θ and variance $-1/E(\partial^2 \log L/\partial\theta^2)$. Thus the relative efficiency of $\hat{\theta}$ tends to 1 as $n \to \infty$, and we therefore say that maximum likelihood estimators are **asymptotically efficient**.

Example 1. A sample of size n is taken from a Poisson distribution with mean θ. The maximum likelihood estimator of θ is the sample mean M, and

$$-E(\partial^2 \log L/\partial\theta^2) = n/\theta.$$

Thus $n^{\frac{1}{2}}(M-\theta)/\theta^{\frac{1}{2}}$ converges in distribution to $\mathcal{N}(0, 1)$, a result which also follows from the central limit theorem.

Example 2. A sample of size n is taken from a normal distribution with mean θ and variance 1. The maximum likelihood estimator of θ is the sample mean M, and

$$-E(\partial^2 \log L/\partial\theta^2) = n.$$

Thus the theorem implies that $n^{\frac{1}{2}}(M-\theta)$ converges in distribution to $\mathcal{N}(0, 1)$, whereas $n^{\frac{1}{2}}(M-\theta)$ is exactly $\mathcal{N}(0,1)$ for all n.

EXERCISES 10

1. The sequence $\{A_{n_1}\}$ converges in probability to a, and the sequence $\{B_{n_2}\}$ converges in probability to b, where a and b are positive. Show that

$$p(|A_{n_1}+B_{n_2}-a-b| > \delta+\varepsilon) \leqslant p(|A_{n_1}-a| > \delta)+p(|B_{n_2}-b| > \varepsilon).$$

Deduce that $(A_{n_1}+B_{n_2})/(a+b)$ converges in probability to 1 as $n_1, n_2 \to \infty$. Prove that $(A_{n_1}/n_1+B_{n_2}/n_2)/(a/n_1+b/n_2)$ also converges in probability to 1.

2. The fact that $n^{\frac{1}{2}}(M-\mu)/S$ converges in distribution to $\mathcal{N}(0, 1)$ for samples from any population with a finite fourth central moment implies that 'Student's' t-distribution tends to normality as the sample

size increases. Denote by $s_\nu(t)$ the density function of the t-distribution with ν degrees of freedom, given in exercise 8.4. Prove that

$$\lim_{\nu \to \infty} s_\nu(t) = \phi(t).$$

3. When X has a Poisson distribution with parameter λ, prove that T_n has a Poisson distribution with parameter $n\lambda$. Apply Stirling's formula to show that

$$p(T_n = j) \sim (2\pi n\lambda)^{-\frac{1}{2}} \exp\{-(j-n\lambda)^2/2n\lambda\}.$$

Compare the exact values of $p(T_n \leqslant j)$ with the approximation

$$p(T_n \leqslant j) \simeq \Phi\{(j+\tfrac{1}{2}-n\lambda)/(n\lambda)^{\frac{1}{2}}\}$$

for $n\lambda = 60$ and $j = 45, 50, \ldots, 75$.

4. An electoral constituency contains n voters, n_1 of whom are Conservative and n_2 Labour. In a random sample of r voters, r_1 are Conservative and r_2 Labour. The ratio $\theta = r/n$ is called the **sampling fraction.** We know n, r_1 and r_2. The problem is to estimate $\psi = n_1/n$. Show that $\hat{\psi} = r_1/r$ is approximately normal with mean ψ and variance $\psi(1-\psi)\theta(1-\theta)/r$. Derive from this result a confidence interval for ψ.

5. The distribution function is $F(x)$, and X_{kn} is the kth order statistic in a sample of size n. Prove that $2nF(X_{kn})$ converges in distribution to χ^2_{2k}.

CHAPTER 11

ANALYSIS OF FREQUENCIES

§52. Binomial variates

In chapter 2 we introduced the more important discrete distributions and showed that they can be used to describe the variation among data consisting of frequencies. Problems which then arose were concerned with how to estimate the unknown parameters and how to test whether the agreement between observed and expected frequencies was satisfactory. The object of this chapter is to develop suitable methods of analysis and describe other techniques which are appropriate for variates with discrete distributions.

We begin with procedures for the analysis of observations on binomial variates. In an experiment which consists of n independent trials with probability θ of success, the number of successes, say B, has a binomial distribution with index n and parameter θ. Thus

$$p(B = b) = \binom{n}{b} \theta^b (1-\theta)^{n-b} \qquad (b = 0, 1, 2, \ldots, n).$$

Consider two such experiments, the first with n_1 trials and probability θ_1 of success, and the second with n_2 trials and probability θ_2 of success. Denote the frequencies of success by B_1 and B_2 respectively. Write

$$B = B_1 + B_2 \quad \text{and} \quad n = n_1 + n_2.$$

The data can be presented as an array of frequencies with two rows and two columns, known as a 2×2 **contingency table**. In the table below, the rows refer to the classifications of success and failure, while the columns refer to experiments 1 and 2.

	Experiments 1	2	*Totals*
Success	B_1	B_2	B
Failure	n_1-B_1	n_2-B_2	$n-B$
Totals	n_1	n_2	n

The values of n_1 and n_2 are given, but θ_1 and θ_2 are unknown. We require to test the hypothesis that $\theta_1 = \theta_2$.

Example 1. The following data are quoted by Cochran (1954). They relate to two groups of school children who came second in birth order. The first group presented behavioural problems, and the second was a comparable group of children with no behavioural problems, described as the controls. For each child, a record was made of whether or not the mother had suffered any infant losses previous to the birth of the child. Here the two groups correspond to experiments 1 and 2, while success denotes no previous losses. The data suggest a higher probability of loss in the problem group.

	Problems	*Controls*	*Totals*
No losses	82	54	136
Losses	20	10	30
Totals	102	64	166

A natural approach to the problem of testing that $\theta_1 = \theta_2$ is as follows. If n_1 is large, the central limit theorem implies that B_1 is approximately normal with mean $n_1\theta_1$ and variance $n_1\theta_1(1-\theta_1)$. Similar remarks apply to B_2. Suppose that $\theta_1 = \theta_2$, and denote the common value by θ. If n_1 and n_2 are large, $B_1/n_1 - B_2/n_2$ is approximately normal with

$$E(B_1/n_1 - B_2/n_2) = 0,$$
$$\text{var}\,(B_1/n_1 - B_2/n_2) = \theta(1-\theta)(1/n_1 + 1/n_2).$$

Consequently the variate

$$U = (B_1/n_1 - B_2/n_2)/\{\theta(1-\theta)(1/n_1 + 1/n_2)\}^{\frac{1}{2}}$$

is approximately $\mathcal{N}(0, 1)$. When the two experiments are combined we have B successes in n trials, so that an unbiased estimator of θ is B/n. This has variance $\theta(1-\theta)/(n_1+n_2)$ and therefore converges in probability to θ as n_1 and n_2 tend to infinity. Substitute B/n for θ in the denominator of U, and we obtain

$$X = \left(\frac{B_1}{n_1} - \frac{B_2}{n_2}\right) \Bigg/ \left\{\frac{B}{n}\left(1 - \frac{B}{n}\right)\left(\frac{1}{n_1} + \frac{1}{n_2}\right)\right\}^{\frac{1}{2}}.$$

Now $(B/n)(1-B/n)$ converges in probability to $\theta(1-\theta)$ by Slutsky's theorem, and therefore X has the same asymptotic distribution as U by Cramér's theorem. Thus X is approximately $\mathcal{N}(0, 1)$ when n_1 and n_2 are large, in which case X^2 is approximately χ_1^2. When $\theta_1 \neq \theta_2$, the absolute value of X will tend to be larger than it is when $\theta_1 = \theta_2$, and therefore large values of X^2 suggest rejection of the hypothesis that $\theta_1 = \theta_2$.

An alternative expression for X^2 can be derived by substituting $\{B_1(n_2-B_2)-B_2(n_1-B_1)\}/n_1n_2$ for $B_1/n_1 - B_2/n_2$. The result is that

$$X^2 = n\{B_1(n_2-B_2)-B_2(n_1-B_1)\}^2/n_1n_2B(n-B).$$

A reference to the 2×2 contingency table set out above shows that $B_1(n_2-B_2)-B_2(n_1-B_1)$ is the cross-product of the frequencies in the body of the table, while $n_1n_2B(n-B)$ is the product of the marginal frequencies for rows and columns. The final expression for X^2 is thus in a form which can be easily remembered and computed.

Example 2. For the data given in example 1, the observed value of X^2 is

$$166(820-1080)^2/(102)(64)(136)(30) = 0{\cdot}421.$$

Since $\chi^2_{1;0\cdot50} = 0{\cdot}455$, this value is not significant. We therefore find that the apparent difference between the two groups is not confirmed.

The test procedure described above can be extended as follows. Suppose that we have k experiments, where the ith experiment consists of n_i independent trials with probability θ_i for success and results in B_i successes $(i = 1, 2, \ldots, k)$. Write

$$B = \Sigma B_i \quad \text{and} \quad n = \Sigma n_i.$$

The data can be presented as an array with 2 rows and k columns, known as a $2xk$ contingency table.

	Experiments						*Totals*
	1	2				k	
Success	B_1	B_2	.	.	.	B_k	B
Failure	n_1-B_1	n_2-B_2	.	.	.	n_k-B_k	$n-B$
Totals	n_1	n_2	.	.	.	n_k	n

Here $n_1, n_2, \ldots, n_k$ are given but $\theta_1, \theta_2, \ldots, \theta_k$ are unknown. The problem is to test the hypothesis

$$\mathscr{H}: \theta_1 = \theta_2 = \ldots = \theta_k.$$

As before, we begin by assuming that the hypothesis is true. Let θ denote the common value of $\theta_1, \theta_2, \ldots, \theta_k$. Consider the device of assigning the value 1 to a trial which results in success and 0 to a trial which results in failure. Then the ith experiment provides a sample of n_i observations and the sample mean is B_i/n_i. Expressed in these terms, the problem is analogous to the one of testing that $\mu_1 = \mu_2 = \ldots = \mu_k$ in §39. The connection between the previous and present notations is that

$$M_i = B_i/n_i, \; M = B/n, \; \text{ and } \; \sigma^2 = \theta(1-\theta).$$

Corresponding to the sum of squares for testing that $\mu_1 = \mu_2 = \ldots = \mu_k$, we now have

$$Q = \Sigma n_i(B_i/n_i)^2 - n(B/n)^2 = \Sigma n_i(B_i/n_i - \theta)^2 - n(B/n-\theta)^2.$$

The variate $n_i^{\frac{1}{2}}(B_i/n_i - \theta)/\{\theta(1-\theta)\}^{\frac{1}{2}}$ converges in distribution to $\mathscr{N}(0, 1)$ as $n_i \to \infty$ for $i = 1, 2, \ldots, k$, and consequently $Q/\theta(1-\theta)$ converges in distribution to χ^2_{k-1}. Substitute B/n for θ, and we obtain the test statistic

$$X^2 = \{\Sigma(B_i^2/n_i) - B^2/n\}/\{B(n-B)/n^2\}.$$

The denominator converges in probability to $\theta(1-\theta)$. Therefore X^2 is approximately χ^2_{k-1} when $n_1, n_2, \ldots, n_k$ are large and $\mathscr{H}$ is true.

§53. Poisson variates

This section is concerned with procedures for the analysis of observations on Poisson variates. Denote by $\mathscr{P}(\lambda)$ a Poisson distribution with mean λ, or a variate having this distribution. We begin by proving that if R is distributed as $\mathscr{P}(\lambda)$ then $(R-\lambda)/\lambda^{\frac{1}{2}}$ converges in distribution to $\mathscr{N}(0, 1)$ as $\lambda \to \infty$. The moment generating function of R is $\exp\{\lambda(e^z - 1)\}$. Hence the moment generating function of $(R-\lambda)/\lambda^{\frac{1}{2}}$, say $J(z)$, is given by

$$\begin{aligned} J(z) &= \exp[-\lambda^{\frac{1}{2}}z + \lambda\{\exp(z/\lambda^{\frac{1}{2}}) - 1\}] \\ &= \exp\{\tfrac{1}{2}z^2 + 0(\lambda^{-\frac{1}{2}})\}. \end{aligned}$$

As $\lambda \to \infty$, $J(z) \to \exp(\frac{1}{2}z^2)$, the moment generating function of $\mathscr{N}(0, 1)$. The same result holds for the characteristic function of

$(R-\lambda)/\lambda^{\frac{1}{2}}$, and the Lévy-Cramér theorem then implies that $(R-\lambda)/\lambda^{\frac{1}{2}}$ converges in distribution to $\mathcal{N}(0, 1)$ as $\lambda \to \infty$.

An immediate application of the result is to determine approximate confidence limits for λ when we have an observation r on a variate R distributed as $\mathcal{P}(\lambda)$. For large λ,

$$p\{(R-\lambda)^2/\lambda < u_{\frac{1}{2}\varepsilon}^2\} \simeq 1-\varepsilon,$$

where $u_{\frac{1}{2}\varepsilon}$ is the upper 50ε percent point of $\mathcal{N}(0, 1)$. The inequality $(R-\lambda)^2 < \lambda u_{\frac{1}{2}\varepsilon}^2$ is equivalent to $f(\lambda) < 0$, where

$$f(\lambda) = \lambda^2 - \lambda(2R + u_{\frac{1}{2}\varepsilon}^2) + R^2.$$

Thus the values of λ which satisfy the inequality lie between the roots of $f(\lambda) = 0$, which are

$$R + \tfrac{1}{2}u_{\frac{1}{2}\varepsilon}^2 \pm u_{\frac{1}{2}\varepsilon}(R + \tfrac{1}{4}u_{\frac{1}{2}\varepsilon}^2)^{\frac{1}{2}}.$$

This shows that $r + \frac{1}{2}u_{\frac{1}{2}\varepsilon}^2 \pm u_{\frac{1}{2}\varepsilon}(r + \frac{1}{4}u_{\frac{1}{2}\varepsilon}^2)^{\frac{1}{2}}$ is a confidence interval for λ with confidence coefficient approximately $1-\varepsilon$. Making a further approximation for large r, we obtain the interval $r \pm u_{\frac{1}{2}\varepsilon} r^{\frac{1}{2}}$.

Example 1. All the cases of acute lymphatic leukaemia reported to the British National Cancer Registration Scheme during 1946–60 were classified by month of clinical onset (Lee, 1963), and 58 cases occurred in June. Assuming that the June frequency is an observation on a Poisson variate, we require confidence limits for λ, the corresponding parameter. Take $\varepsilon = 0{\cdot}05$ so that $u_{\frac{1}{2}\varepsilon} = 1{\cdot}96$. A confidence interval for λ with confidence coefficient approximately 95 percent is

$$58 + 1{\cdot}92 \pm (58 + 0{\cdot}96)^{\frac{1}{2}} = 59{\cdot}92 \pm 7{\cdot}68.$$

Suppose that $R_1, R_2, \ldots, R_k$ are independent variates distributed as $\mathcal{P}(\lambda_1), \mathcal{P}(\lambda_2), \ldots, \mathcal{P}(\lambda_k)$ respectively. We shall need the joint distribution of $R_1, R_2, \ldots, R_k$ conditional on $\Sigma R_i = r$. The unconditional joint distribution is

$$p(R_1 = r_1, R_2 = r_2, \ldots, R_k = r_k) = \prod_{i=1}^{k} (\lambda_i^{r_i} e^{-\lambda_i}/r_i!)$$

$$= e^{-\Sigma\lambda_i} \prod_{i=1}^{k} (\lambda_i^{r_i}/r_i!).$$

Now ΣR_i is $\mathcal{P}(\Sigma\lambda_i)$ because the probability generating function of

R_i is exp $\{\lambda_i(s-1)\}$, whence the probability generating function of ΣR_i is exp $\{\Sigma\lambda_i(s-1)\}$. Thus

$$p(\Sigma R_i = r) = (\Sigma\lambda_i)^r e^{-\Sigma\lambda_i}/r!.$$

Take $r = \Sigma r_i$ and write θ_i for $\lambda_i/\Sigma\lambda_i$. We then have

$$p(R_1 = r_1, R_2 = r_2, \ldots, R_k = r_k \mid \Sigma R_i = r) = r! \prod_{i=1}^{k} (\theta_i^{r_i}/r_i!),$$

which is a multinomial distribution with index r and parameters $\theta_1, \theta_2, \ldots, \theta_k$, where $\Sigma\theta_i = 1$.

This is a basic result for the arguments of the next section and can also be applied as follows. Suppose that observations are made on independent variates $R_1, R_2, \ldots, R_k$ which are distributed as $\mathscr{P}(\lambda_1), \mathscr{P}(\lambda_2), \ldots, \mathscr{P}(\lambda_k)$ respectively. The problem is to test the hypothesis

$$\mathscr{H}: \lambda_1 = \lambda_2 = \ldots = \lambda_k.$$

When $\mathscr{H}$ is true, denote the unknown common value of the parameters by λ, so that $R_1, R_2, \ldots, R_k$ correspond to a sample of size k from $\mathscr{P}(\lambda)$. The expectation of the sample mean and sample variance both equal λ, which suggests that a possible statistic for testing $\mathscr{H}$ is the ratio of the sample variance to the sample mean. An equivalent test statistic is

$$D = \sum_{i=1}^{k} (R_i - M)^2/M,$$

where M is the sample mean, $\Sigma R_i/k$.

We proceed to find $E(D)$, using the expression

$$E(D) = E\{E(D \mid M)\},$$

derived in §**20**. The joint distribution of $R_1, R_2, \ldots, R_k$ conditional on $kM = r$ is multinomial with index r and parameters $\theta_1, \theta_2, \ldots, \theta_k$ where $\theta_i = \lambda_i/\Sigma\lambda_i$. Therefore the conditional distribution of R_i is binomial with index r and parameter θ_i, so that

$$E(R_i^2 \mid kM = r) = r\theta_i(1-\theta_i) + r^2\theta_i^2.$$

When $M \neq 0$, D can be written in the form $\Sigma R_i^2/M - kM$, and hence

$$\begin{aligned} E(D \mid kM = r) &= k\Sigma\theta_i(1-\theta_i) + kr\Sigma\theta_i^2 - r \\ &= k + k(r-1)\Sigma\theta_i^2 - r. \end{aligned}$$

When $M = 0$, we define D as $k(1-\Sigma\theta_i^2)$ for consistency. Since kM is $\mathscr{P}(\Sigma\lambda_i)$,

$$E(D) = k+k(\Sigma\lambda_i-1)\Sigma\theta_i^2-\Sigma\lambda_i.$$

Now fix $\Sigma\lambda_i$ at a value which exceeds 1, and minimize $E(D)$ with respect to $\theta_1, \theta_2, \ldots, \theta_k$. The condition $\Sigma\theta_i = 1$ implies that

$$\Sigma\theta_i^2 \equiv \Sigma(\theta_i-1/k)^2+1/k.$$

Hence $E(D)$ has a minimum value of $(k-1)$ which is attained when $\theta_i = 1/k$ for all i, in other words, when $\mathscr{H}$ is true. We infer that large values of D should be regarded as significant when $\mathscr{H}$ is tested.

Suppose that $R_1, R_2, \ldots, R_k$ are independent variates each distributed as $\mathscr{P}(\lambda)$. We shall prove that D converges in distribution to χ^2_{k-1} as $\lambda \to \infty$. Denote $(R_i-\lambda)/\lambda^{\frac{1}{2}}$ by Y_i. As $\lambda \to \infty$, Y_i converges in distribution to Z_i, where Z_i is $\mathscr{N}(0, 1)$. A consequence of Sverdrup's theorem is that $\Sigma(Y_i-\Sigma Y_i/k)^2$ converges in distribution to $\Sigma(Z_i-\Sigma Z_i/k)^2$. We have shown in §**38** that $\Sigma(Z_i-\Sigma Z_i/k)^2$ is distributed as χ^2_{k-1}. Therefore $\Sigma(Y_i-\Sigma Y_i/k)^2$ converges in distribution to χ^2_{k-1}, and the same asymptotic distribution holds for $\Sigma(R_i-M)^2/\lambda$ because

$$\Sigma(Y_i-\Sigma Y_i/k)^2 = \Sigma(R_i-M)^2/\lambda.$$

We continue by proving that M/λ converges in probability to 1. In fact,

$$E\{(M/\lambda-1)^2\} = \text{var}\,(M)/\lambda^2 = 1/k\lambda,$$

which tends to zero as $\lambda \to \infty$. Thus M/λ converges in mean square to 1, and hence also in probability to 1. Since D is the ratio of $\Sigma(R_i-M^2)/\lambda$ to M/λ, Cramér's theorem implies that D converges in distribution to χ^2_{k-1} as $\lambda \to \infty$.

Example 2. In example 1, we referred to a monthly classification of leukaemia cases, and the frequencies for all months are given below.

Jan.	*Feb.*	*Mar.*	*Apr.*	*May*	*June*	*July*	*Aug.*	*Sept.*	*Oct.*	*Nov.*	*Dec.*
40	34	30	44	39	58	51	55	36	48	33	38

Assuming that each frequency is an observation on a Poisson variate we wish to test that the corresponding parameters $\lambda_1, \lambda_2, \ldots, \lambda_{12}$

are equal. Here $D = 21{\cdot}1$, as compared with $\chi^2_{11;0\cdot05} = 19{\cdot}7$, which suggests a variation of λ_i between months.

D is known as the **Poisson index of dispersion**. The meaning of this term can be illustrated by reference to a situation where objects are distributed in space, for example, plants in a field. Divide the field into k parts of equal area, and denote the corresponding frequencies of plants by $R_1, R_2, \ldots, R_k$. If the plants are distributed independently and at random, then $R_1, R_2, \ldots, R_k$ are independently and identically distributed as Poisson variates, so that $E(D) = (k-1)$. If the plants form a regular pattern, then the frequencies in the different parts are roughly equal and D is small or zero. If the plants form clusters, then some values of R_i are much larger than average and others much smaller, so that D tends to be large. Thus D is an index of dispersion in the sense that various types of distribution are associated with a wide range of values of D.

§54. Exact test procedures

The test procedures described in the two previous sections are based on asymptotic distributions and are therefore valid only as approximations when frequencies are large. Such procedures cannot justifiably be used when frequencies are small without a further study of their properties, for example, by simulating a large number of observations and estimating the probability distribution of the test statistic. We now show that test procedures based on exact distributions can be found in some problems of frequency-analysis.

Consider again the first problem treated in §52. We have two experiments, each of which consists of a sequence of independent trials classified as success or failure, and require to test that the probability of success is the same for each experiment. The notation will be the same as before. In addition, b_i denotes the observed value of B_i $(i = 1, 2)$, $b = b_1+b_2$, and $B = B_1+B_2$. Suppose that θ_1 and θ_2 have a common value θ. Since B_1 and B_2 are independent,

$$p(B_1 = b_1, B_2 = b_2) = \binom{n_1}{b_1}\binom{n_2}{b_2}\theta^b(1-\theta)^{n-b}.$$

This shows that B is sufficient for θ. According to example 1 of §22, the probability distribution of B is

$$p(B = b) = \binom{n}{b}\theta^b(1-\theta)^{n-b}.$$

We therefore obtain

$$p(B_1 = b_1 \mid B = b) = \binom{n_1}{b_1}\binom{n_2}{b_2}\Big/\binom{n}{b},$$

which shows that B_1 has a hypergeometric distribution when $\theta_1 = \theta_2$ and B is fixed. As noted in §**8**, there are non-zero probabilities for

$$\max(0, b-n_2) \leqslant b_1 \leqslant \min(n_1, b).$$

Since B_1+B_2 is fixed, values of b_1 near their maximum are associated with values of b_2 near their minimum, and consequently with the largest positive values of $b_1/n_1-b_2/n_2$. Similarly, values of b_1 near their minimum are associated with the largest negative values of $b_1/n_1-b_2/n_2$. Thus the values of b_1 in the tails of the hypergeometric distribution correspond to the largest values of $|b_1/n_1-b_2/n_2|$ and therefore suggest rejection of the hypothesis that $\theta_1 = \theta_2$.

A formal procedure of testing $\mathscr{H}: \theta_1 = \theta_2$ against the two-sided alternative $\theta_1 \neq \theta_2$ is defined as follows. We begin by selecting a small probability ε. If $b_1/n_1 > b_2/n_2$, we reject $\mathscr{H}$ whenever

$$p(B_1 \geqslant b_1 \mid B = b) \leqslant \tfrac{1}{2}\varepsilon.$$

Notice that the probability can seldom equal $\frac{1}{2}\varepsilon$ because of the discreteness of the hypergeometric distribution. If $b_1/n_1 < b_2/n_2$, we similarly reject $\mathscr{H}$ whenever

$$p(B_1 \leqslant b_1 \mid B = b) \leqslant \tfrac{1}{2}\varepsilon.$$

Rejection does not arise if $b_1/n_1 = b_2/n_2$, and thus the probability that $\mathscr{H}$ is rejected when $B = b$ is at most ε. On taking an expectation over B, we conclude that the unconditional probability of rejection is at most ε.

When n_1 and n_2 are 15 or less, values of b_1 such that

$$p(B_1 \leqslant b_1 \mid B = b) \leqslant \tfrac{1}{2}\varepsilon,$$

and the exact probabilities of rejection are given by Pearson and Hartley (1966) for various values of ε. The probabilities of the hypergeometric distribution can, if required, be calculated from the expression

$$p(B_1 = b_1 \mid B = b) \\ = n_1!\, n_2!\, b!\, (n-b)!\, /\, n!\, b_1!\, (n_1-b_1)!\, b_2!\, (n_2-b_2)!$$

using a table of log factorials.

Example. The incidence of pathogenic *E. coli* infection among newborn babies with congenital abnormalities of the large intestine was compared with the incidence among all other groups by Lapatsanis and Irving (1963). Their results are as follows:

	With symptoms	*Without symptoms*	*Totals*
Abnormal group	8	4	12
Other groups	14	28	42
Totals	22	32	54

We require to test that the proportion with symptoms is the same for the abnormal group as for all other groups, against the alternative hypothesis that the proportion is higher for the abnormal group. The hypergeometric probability corresponding to the table observed is 0·03352. Each of the tables below has the same marginal frequencies as the table observed, and the corresponding probabilities are appended.

9	3
13	29

0·00719

10	2
12	30

0·00094

11	1
11	31

0·00007

12	0
10	32

0·00000

The total probability for the five tables is 0·04172. There is some evidence that the proportion with symptoms is higher in the abnormal group.

When n_1, n_2, b and $n-b$ all tend to infinity such that $n_1/n \to \psi$ and $b/n \to \theta$, we have proved in §**49** that

$$(B_1 - n\psi\theta)/\{n\psi(1-\psi)\theta(1-\theta)\}^{\frac{1}{2}}$$

converges in distribution to $\mathcal{N}(0, 1)$. For large n, the procedure of testing $\mathcal{H}$ against two-sided alternatives therefore consists in referring

$$(b_1 - n_1 b/n)^2/\{n_1 n_2 b(n-b)/n^3\}$$
$$= n\{b_1(n_2-b_2) - b_2(n_1-b_1)\}^2/n_1 n_2 b(n-b)$$

to the distribution of χ_1^2. This is exactly the procedure derived in §**52** for the same problem, so that the two methods of testing $\mathcal{H}$ coincide in large samples.

Another illustration of an exact test procedure is as follows. The variates R_1 and R_2 are independently distributed as $\mathcal{P}(\lambda_1)$ and $\mathcal{P}(\lambda_2)$

respectively. We require to test the hypothesis that $\lambda_1 = \lambda_2$. Denote the observed values of R_1 and R_2 by r_1 and r_2 respectively, and write $r = r_1+r_2$. The distribution of R_1 conditional on $R_1+R_2 = r$ is a special case of the multinomial distribution derived in the previous section. We obtain

$$p(R_1 = r_1 \mid R_1+R_2 = r) = \binom{r}{r_1}\left(\frac{\lambda_1}{\lambda_1+\lambda_2}\right)^{r_1}\left(1-\frac{\lambda_1}{\lambda_1+\lambda_2}\right)^{r-r_1},$$

a binomial distribution with index r and parameter $\lambda_1/(\lambda_1+\lambda_2)$.

On the hypothesis $\mathscr{L}: \lambda_1 = \lambda_2$, the value of $\lambda_1/(\lambda_1+\lambda_2)$ is $\frac{1}{2}$. The procedure for testing $\mathscr{L}$ is thus based on a binomial distribution with index r and parameter $\frac{1}{2}$. Suppose that the alternative hypothesis is $\lambda_1 \neq \lambda_2$. If $r_1 > r_2$, we reject $\mathscr{L}$ whenever

$$2^{-r} \sum_{i=r_1}^{r} \binom{r}{i} \leqslant \tfrac{1}{2}\varepsilon.$$

Similarly, if $r_1 < r_2$, we reject whenever

$$2^{-r} \sum_{i=0}^{r_1} \binom{r}{i} \leqslant \tfrac{1}{2}\varepsilon.$$

The probability of rejection is at most ε conditional on $R_1+R_2 = r$, and is therefore at most ε unconditionally.

§55. Multinomial variates with specified probabilities

Consider an experiment consisting of n independent trials, at each of which there are k possible outcomes. Denote the frequencies of the different outcomes by $N_1, N_2, \ldots, N_k$ and let the corresponding probabilities be $p_1, p_2, \ldots, p_k$ respectively. Thus

$$\Sigma N_i = n \quad \text{and} \quad \Sigma p_i = 1.$$

As shown in §**10**, $N_1, N_2, \ldots, N_k$ follow a multinomial distribution

$$p(N_1 = n_1, N_2 = n_2, \ldots, N_k = n_k)$$
$$= n!\, p_1^{n_1} p_2^{n_2} \ldots p_k^{n_k}/n_1!\, n_2! \ldots n_k!\,.$$

We shall discuss the problem of testing that the observed frequencies $n_1, n_2, \ldots, n_k$ are reasonably consistent with particular values of $p_1, p_2, \ldots, p_k$.

Example 1. The data below are taken from the Liverpool telephone directory for 1968, p. 1500, column 1. They are the frequencies with which 0, 1, 2, . . . , 9 occur as the last digit among the 100 numbers listed.

Digit	0	1	2	3	4	5	6	7	8	9
Frequency	15	8	11	7	10	14	5	6	13	11

Let p_i be the probability for the digit $(i-1)$. On the hypothesis that there is no association between names and final digits, $p_i = 0{\cdot}1$ for $i = 1, 2, \ldots, 10$.

Example 2. The data in Table 4 are the frequencies with which all pairs of digits are observed when two random decimal digit generators are operated simultaneously. There are 100 pairs and each of the corresponding probabilities is 0·01 on the hypothesis that the generators are truly random and the digits of a pair are independent.

In general, suppose that we wish to test the hypothesis

$$\mathscr{H}: p_i = \pi_i \qquad (i = 1, 2, \ldots, k)$$

where $\pi_1, \pi_2, \ldots, \pi_k$ are a set of specified values such that $\Sigma\pi_i = 1$. The test statistic

$$L^2 = \sum_{i=1}^{k} (N_i - n\pi_i)^2/n\pi_i$$

was proposed by Karl Pearson. Some justification for the use of L^2 to test $\mathscr{H}$ is provided by considering $E(L^2)$. For any variate N_i,

$$E(N_i - n\pi_i)^2 = \text{var}\,(N_i - n\pi_i) + \{E(N_i - n\pi_i)\}^2.$$

Here N_i has a binomial distribution with index n and parameter p_i, whence

$$E(N_i - n\pi_i)^2 = np_i(1-p_i) + n^2(p_i - \pi_i)^2.$$

We therefore obtain

$$E(L^2) = \sum_{i=1}^{k} \{p_i(1-p_i)/\pi_i\} + n \sum_{i=1}^{k} (p_i - \pi_i)^2/\pi_i.$$

If $\mathscr{H}$ is true, then $E(L^2) = k-1$, but otherwise $E(L^2) \to \infty$ as $n \to \infty$. Thus a large observed value of L^2 suggests a departure from $\mathscr{H}$.

We now show that L^2 converges in distribution to χ^2_{k-1} as $n \to \infty$ when $\mathscr{H}$ is true. Suppose that $N_1, N_2, \ldots, N_k$ are independently distributed as $\mathscr{P}(n\pi_1), \mathscr{P}(n\pi_2), \ldots, \mathscr{P}(n\pi_k)$ respectively. According to a result in the previous section, the joint distribution of

$N_1, N_2, \ldots, N_k$ conditional on $\Sigma N_i = n$ is a multinomial distribution with index n and parameters $n\pi_i/\Sigma n\pi_i = \pi_i$ for $i = 1, 2, \ldots, k$. Write

$$Y_i = (N_i - n\pi_i)/(n\pi_i)^{\frac{1}{2}}.$$

When N_i is $\mathscr{P}(n\pi_i)$, Y_i converges in distribution to $\mathscr{N}(0, 1)$ as $n \to \infty$. Make an orthonormal transformation from $Y_1, Y_2, \ldots, Y_k$ to $T_1, T_2, \ldots, T_k$ such that

$$T_1 = \sum_{i=1}^{k} Y_i \pi_i^{\frac{1}{2}},$$

possible because $\Sigma(\pi_i^{\frac{1}{2}})^2 = 1$. Thus

$$T_1 = \sum_{i=1}^{k} (N_i - n\pi_i)/n^{\frac{1}{2}}.$$

Now $L^2 = \Sigma_1^k Y_i^2 = \sum_1^k T_j^2$. The condition $\Sigma N_i = n$ implies that $T_1 = 0$ and $L^2 = \Sigma_2^k T_j^2$. Since $\Sigma_2^k T_j^2$ is asymptotically χ^2_{k-1}, we have proved the following result. If $N_1, N_2, \ldots, N_k$ have a multinomial distribution with index n and parameters $\pi_1, \pi_2, \ldots, \pi_k$ respectively, then $\Sigma(N_i - n\pi_i)^2/n\pi_i$ converges in distribution to χ^2_{k-1} as $n \to \infty$.

The fact that

$$\Sigma(N_i - n\pi_i)^2/n\pi_i = \Sigma N_i^2/n\pi_i - 2\Sigma N_i + n\Sigma\pi_i = \Sigma N_i^2/n\pi_i - n$$

shows that we have two expressions for L^2, namely

$$\sum_{i=1}^{k} (N_i - n\pi_i)^2/n\pi_i \quad \text{and} \quad \sum_{i=1}^{k} N_i^2/n\pi_i - n.$$

Although the second version is easier to calculate, the first is preferable because an inspection of the differences $\{N_i - n\pi_i\}$ may suggest other hypotheses when L^2 is significantly large.

Example 3. The frequencies with which 0, 1, 2, . . . , 9 occur as the last digit in a sample of telephone numbers are given in example 1. Here the observed value of L^2 is

$$1106/10 - 100 = 10{\cdot}6,$$

which is close to $\chi^2_{9;0\cdot30} = 10{\cdot}66$. We accept the hypothesis that each final digit has the same probability.

§56. Multinomial variates with unknown probabilities

We seldom have exact knowledge of $p_1, p_2, \ldots, p_k$. These probabilities are usually functions of unknown parameters, and the first problem is to estimate the parameters from the data. We suppose throughout this section that there is a single unknown parameter θ, and that the multinomial probabilities are given by continuous functions $p_1(\theta), p_2(\theta), \ldots, p_k(\theta)$. The likelihood of $n_1, n_2, \ldots, n_k$ is

$$l = n!\,\{p_1(\theta)\}^{n_1}\{p_2(\theta)\}^{n_2}\ldots\{p_k(\theta)\}^{n_k}/n_1!\,n_2!\ldots n_k!\,.$$

Denote the corresponding variate by L. The maximum likelihood estimator $\hat{\theta}$ satisfies $\partial \log L/\partial\theta = 0$, which takes the form

$$\sum_{i=1}^{k} N_i\{\partial \log p_i(\theta)/\partial\theta\} = 0.$$

Example 1. In Table 1 (p. 19) we gave the observed frequencies of boys among the first 5 children in 2571 families of size 5 or more. By assumption, the probability that a family of size 5 contains j boys is

$$p_{j+1}(\theta) = \binom{5}{j}\theta^j(1-\theta)^{5-j} \qquad (j = 0, 1, \ldots, 5).$$

The corresponding frequency is N_{j+1}. We require

$$\frac{\partial}{\partial\theta}\{\log p_{j+1}(\theta)\} = \frac{j}{\theta} - \frac{5-j}{1-\theta} = \frac{j-5\theta}{\theta(1-\theta)}.$$

The estimator $\hat{\theta}$ satisfies

$$\sum_{j=0}^{5} N_{j+1}(j-5\hat{\theta})/\hat{\theta}(1-\hat{\theta}) = 0,$$

from which we obtain

$$\hat{\theta} = \sum_{j=1}^{5} j\,N_{j+1}/5n.$$

This is precisely the estimator used to calculate the expected frequencies in Table 1.

The value of $E(\partial^2 \log L/\partial\theta^2)$ is required in order to apply the limit theorem for maximum likelihood estimators. Since $E(N_i) = np_i(\theta)$ we obtain

$$E\left(\frac{\partial^2 \log L}{\partial\theta^2}\right) = n\sum_{i=1}^{k} p_i(\theta)\,\frac{\partial^2 \log p_i(\theta)}{\partial\theta^2}.$$

By an argument given in §**32**,

$$\frac{\partial^2 \log p_i}{\partial\theta^2} = \frac{1}{p_i}\frac{\partial^2 p_i}{\partial\theta^2} - \frac{1}{p_i^2}\left(\frac{\partial p_i}{\partial\theta}\right)^2.$$

The second derivative of Σp_i is zero, and consequently an alternative expression is

$$E\left(\frac{\partial^2 \log L}{\partial\theta^2}\right) = -n \sum_{i=1}^{k} p_i(\theta)\left\{\frac{\partial \log p_i(\theta)}{\partial\theta}\right\}^2.$$

We deduce from the limit theorem that $\hat{\theta}$ differs from θ by $0(n^{-\frac{1}{2}})$ with a probability which tends to 1 as $n \to \infty$.

Example 2. The probability that a family of size 5 contains j boys is

$$p_{j+1}(\theta) = \binom{5}{j}\theta^j(1-\theta)^{5-j} \qquad (j = 0, 1, \ldots, 5).$$

From example 1, we find that

$$E(\partial^2 \log L/\partial\theta^2) = -n\Sigma p_{j+1}(\theta)\,(j-5\theta)^2/\theta^2(1-\theta)^2.$$

Now $\Sigma(j-5\theta)^2 p_{j+1}(\theta)$ is the variance of a binomial distribution with index 5 and parameter θ. Hence

$$E(\partial^2 \log L/\partial\theta^2) = -5n/\theta(1-\theta).$$

Thus $(\hat{\theta}-\theta)\{5n/\theta(1-\theta)\}^{\frac{1}{2}}$ converges in distribution to $\mathcal{N}(0, 1)$ as $n \to \infty$.

We turn now to the problem of testing that the frequencies $N_1, N_2, \ldots, N_k$ are consistent with a multinomial distribution for which the total frequency is n and the probabilities are $p_1(\theta), p_2(\theta), \ldots, p_k(\theta)$ respectively. In what follows, θ denotes a parameter which takes values within a specified range, and θ_0 denotes the true value of θ in the experiment. Since $|\hat{\theta}-\theta_0|$ is $0(n^{-\frac{1}{2}})$ with a probability which tends to 1 as $n \to \infty$, we need consider only values of θ such that $|\theta-\theta_0|$ is $0(n^{-\frac{1}{2}})$. Proceeding as before, we suppose that the joint distribution of $N_1, N_2, \ldots, N_k$ is obtained from independent variates distributed respectively as

$$\mathscr{P}(np_1(\theta)\}, \mathscr{P}\{np_2(\theta)\}, \ldots, \mathscr{P}\{np_k(\theta)\},$$

and subject to the condition $\Sigma N_i = n$. Define

$$Y_i = \{N_i - np_i(\theta)\}/\{np_i(\theta)\}^{\frac{1}{2}} \qquad (i = 1, 2, \ldots, k).$$

As $n \to \infty$, Y_i converges in distribution to $\mathcal{N}(0, 1)$. By a change in the denominator of Y_i, we obtain

$$Z_i = \{N_i - np_i(\theta)\}/\{np_i(\theta_0)\}^{\frac{1}{2}} \qquad (i = 1, 2, \ldots, k).$$

The condition that $|\theta - \theta_0|$ is $0(n^{-\frac{1}{2}})$ and the continuity of $p_i(\theta)$ together imply that $|p_i(\theta) - p_i(\theta_0)|$ is $0(n^{-\frac{1}{2}})$. Hence Z_i also converges in distribution to $\mathcal{N}(0, 1)$. Let $\partial p_i(x)/\partial\theta$ be the value of $\partial p_i/\partial\theta$ at $\theta = x$. Write

$$c = \sum_{i=1}^{k} \{p_i(\theta_0)\}^{-1} \left\{\frac{\partial p_i(\theta_0)}{\partial\theta}\right\}^2.$$

Make an orthonormal transformation from $Z_1, Z_2, \ldots, Z_k$ to $T_1, T_2, \ldots, T_k$ where

$$T_1 = \sum_{i=1}^{k} \{p_i(\theta_0)\}^{\frac{1}{2}} Z_i \quad \text{and} \quad T_2 = \sum_{i=1}^{k} \{cp_i(\theta_0)\}^{-\frac{1}{2}} \frac{\partial p_i(\theta_0)}{\partial\theta} Z_i.$$

The coefficients in T_1 and T_2 define orthonormal vectors because

$$\sum_{i=1}^{k} p_i(\theta_0) = 1, \quad \sum_{i=1}^{k} \{cp_i(\theta_0)\}^{-1} \left\{\frac{\partial p_i(\theta_0)}{\partial\theta}\right\}^2 = 1,$$

and
$$\sum_{i=1}^{k} c^{-\frac{1}{2}} \frac{\partial p_i(\theta_0)}{\partial\theta} = c^{-\frac{1}{2}} \frac{\partial}{\partial\theta} \sum_{i=1}^{k} p_i(\theta_0) = 0.$$

Consider $\Sigma_3^k T_j^2$, which is the value of $\Sigma_1^k Z_i^2$ when $T_1 = 0$ and $T_2 = 0$. The condition $T_1 = 0$ implies that $\Sigma N_i = n$ and ensures that $N_1, N_2, \ldots, N_k$ have a multinomial distribution. Now consider the condition $T_2 = 0$, which implies that

$$\sum_{i=1}^{k} \frac{1}{p_i(\theta_0)} \frac{\partial p_i(\theta_0)}{\partial\theta} \{N_i - np_i(\theta)\} = 0.$$

Here $N_i - np_i(\theta)$ is $0(n^{\frac{1}{2}})$ with a probability which tends to 1 as $n \to \infty$, whereas

$$\frac{1}{p_i(\theta_0)} \frac{\partial p_i(\theta_0)}{\partial\theta} \quad \text{and} \quad \frac{1}{p_i(\theta)} \frac{\partial p_i(\theta)}{\partial\theta}$$

differ by $0(n^{-\frac{1}{2}})$. Therefore $T_2 = 0$ is asymptotically equivalent to the condition

$$\sum_{i=1}^{k} \frac{1}{p_i(\theta)} \frac{\partial p_i(\theta)}{\partial\theta} \{N_i - np_i(\theta)\} = 0,$$

which can also be expressed as

$$\sum_{i=1}^{k} N_i \{\partial \log p_i(\theta)/\partial\theta\} = 0,$$

because the first derivative of $\Sigma p_i(\theta)$ is zero. The last equation is satisfied by $\theta = \hat{\theta}$, which also fulfils the requirement that $|\theta-\theta_0|$ is $0(n^{-\frac{1}{2}})$ with a probability which tends to 1 as $n \to \infty$. Consequently

$$\sum_{j=3}^{k} T_j^2 = \sum_{i=1}^{k} \{N_i - np_i(\hat{\theta})\}^2/np_i(\theta_0).$$

Replace θ_0 by $\hat{\theta}$ and we obtain X^2, defined by

$$X^2 = \sum_{i=1}^{k} \{N_i - np_i(\hat{\theta})\}^2/np_i(\hat{\theta}).$$

Since $\Sigma_3^k T_j^2$ is asymptotically χ_{k-2}^2, and $\hat{\theta}$ converges in probability to θ_0, Cramér's theorem implies that X^2 is also asymptotically χ_{k-2}^2. We have thus established the following result. If $N_1, N_2, \ldots, N_k$ have a multinomial distribution with total frequency n and probabilities $p_1(\theta), p_2(\theta), \ldots, p_k(\theta)$ respectively, and if $\hat{\theta}$ is the solution of $\Sigma N_i(\partial \log p_i/\partial\theta) = 0$, then X^2 converges in distribution to χ_{k-2}^2 as $n \to \infty$.

Example 3. The observed value of X^2 for the data given in Table 1 (p. 19) is

$$(63-69{\cdot}18)^2/69{\cdot}18 + \ldots + (85-92{\cdot}91)^2/92{\cdot}91 = 1{\cdot}53.$$

If the data form a sample of 2571 from a multinomial distribution with binomial probabilities for the six groups, then the distribution of X^2 is close to χ_4^2. The observed value of X^2 is near $\chi_{4;0{\cdot}80}^2$ and therefore supports the assumption of binomial probabilities.

Example 4. In §**12** we assumed an exponential distribution for the duration of the interval between emission of α-particles by uranium. The data are given in Table 5 (p. 37), where the duration in sec. is classified into ten groups, namely 0–1, 1–2, . . . , 8–9, and 9–∞. Denote the density function by $\theta e^{-\theta x}$. Then the probability for an interval between $(i-1)$ and i sec. is

$$p_i(\theta) = e^{-(i-1)\theta} - e^{-i\theta} \qquad (i = 1, 2, \ldots, 9)$$

from which we obtain

$$\frac{1}{p_i}\frac{\partial p_i}{\partial\theta} = \frac{-(i-1)e^{-(i-1)\theta} + i\,e^{-i\theta}}{e^{-(i-1)\theta} - e^{-i\theta}} = \frac{1}{1-e^{-\theta}} - i.$$

For the last interval

$$p_{10}(\theta) = e^{-9\theta} \quad \text{and} \quad \partial \log p_{10}/\partial\theta = -9.$$

The equation of estimation for θ is therefore

$$(1-e^{-\hat{\theta}})^{-1} \sum_{i=1}^{9} N_i = \sum_{i=1}^{9} iN_i + 9N_{10}.$$

When the observed values of N_i are inserted,

$$1-e^{-\hat{\theta}} = 7495/18378 = 0{\cdot}4078.$$

Successive multiplications by $e^{-\hat{\theta}}$ give $p_2(\hat{\theta}), \ldots, p_9(\hat{\theta})$. The condition $\Sigma p_i(\hat{\theta}) = 1$ then leads to $p_{10}(\hat{\theta})$. We do not explicitly require $\hat{\theta} = 0{\cdot}52395$, which is the value adopted in §**12**. The observed value of X^2 is 6·93, as compared with $\chi^2_{8;\,0\cdot50} = 7{\cdot}34$. We conclude that an exponential distribution satisfactorily describes the data.

EXERCISES 11

1. An experiment consists of n independent trials with probability θ for success, and the observed number of successes is b. If n is large, show that a confidence interval for θ, with confidence coefficient approximately $1-\varepsilon$, is given by

$$\frac{n}{n+u^2_{\frac{1}{2}\varepsilon}}\left[\frac{b}{n}+\frac{u^2_{\frac{1}{2}\varepsilon}}{2n} \pm u_{\frac{1}{2}\varepsilon}\left\{\frac{b(n-b)}{n^3}+\frac{u^2_{\frac{1}{2}\varepsilon}}{4n^2}\right\}^{\frac{1}{2}}\right].$$

2. The variates R_1 and R_2 are independently distributed as $\mathscr{P}(\lambda_1)$ and $\mathscr{P}(\lambda_2)$ respectively. We can test the hypothesis that $\lambda_1 = \lambda_2$ by two procedures. The first is based on the index of dispersion, and the second on the distribution of R_1 conditional on the observed value of R_1+R_2. When $\lambda_1 = \lambda_2$, denote the common value by λ. In this case, show that the two test procedures are equivalent in the limit as $\lambda \to \infty$.

3. An experiment consists of n independent trials, and the result of each trial is classified into two groups by each of two criteria. Let N_{ij} be the frequency of trials classified as group i by the first criterion and group j by the second ($i, j = 1, 2$). The data can be presented in the form of a 2×2 contingency table where the rows refer to groups formed by one criterion, and the columns refer to groups formed by the other. A zero suffix is used in row and column totals as the table indicates. For example, $N_{10} = N_{11}+N_{12}$.

		Second criterion		
		Group 1	*Group* 2	*Totals*
First criterion	*Group* 1	N_{11}	N_{12}	N_{10}
	Group 2	N_{21}	N_{22}	N_{20}
	Totals	N_{01}	N_{02}	n

Denote the probability corresponding to N_{ij} by π_{ij}. Thus $N_{11}, N_{12}, N_{21}, N_{22}$ have a multinomial distribution with total frequency n and probabilities $\pi_{11}, \pi_{12}, \pi_{21}, \pi_{22}$ respectively. Write $\theta = \pi_{11}+\pi_{12}$ and $\psi = \pi_{11}+\pi_{21}$. The hypothesis that the criteria of classification are independent is expressed by

$$\pi_{11} = \theta\psi,\ \pi_{12} = \theta(1-\psi),\ \pi_{21} = (1-\theta)\psi,\ \pi_{22} = (1-\theta)(1-\psi).$$

When independence holds, prove that

$$p(N_{11} = n_{11} \mid N_{10} = n_{10}, N_{01} = n_{01}) \\ = n_{10}!\, n_{20}!\, n_{01}!\, n_{02}! \,/\, n!\, n_{11}!\, n_{12}!\, n_{21}!\, n_{22}!\,.$$

Show how an exact test procedure for the hypothesis of independence can be based on this conditional distribution.

4. In a sequence of k experiments, the ith experiment consists of n independent trials with probability θ_i for success and results in B_i successes ($i = 1, 2, \ldots, k$). Write $B = \Sigma B_i$ and $n = \Sigma n_i$. When $\theta_1 = \theta_2 = \ldots = \theta_k$, prove that

$$p(B_1 = b_1, B_2 = b_2, \ldots, B_k = b_k \mid B = b) \\ = \frac{b!\,(n-b)!\,n_1!\,n_2!\ldots n_k!}{b_1!\,(n_1-b_1)!\,b_2!\,(n_2-b_2)!\ldots b_k!\,(n_k-b_k)!}\,.$$

By using $\Sigma n_i(B_i/n_i - B/n)^2$ as a measure of departure from the hypothesis that $\theta_1 = \theta_2 = \ldots = \theta_k$, show how an exact test procedure can be based on the conditional distribution.

5. The conditions are the same as in the previous exercise. In addition the ith experiment is associated with a fixed value x_i, and the number of successes in a single trial is Y_i ($i = 1, 2, \ldots, k$). Thus

$$p(Y_i = 1) = \theta_i \quad \text{and} \quad p(Y_i = 0) = 1-\theta_i.$$

Suppose that there is a linear trend in the probability of success, so that

$$\theta_i = \theta+\beta x_i \qquad (i = 1, 2, \ldots, k).$$

We require to test the hypothesis that $\beta = 0$. By analogy with linear regression, show that an appropriate test statistic is T^2, defined by

$$T^2 = \frac{n^2(\Sigma x_i B_i - B\Sigma n_i x_i/n)^2}{B(n-B)\{\Sigma n_i x_i^2 - (\Sigma n_i x_i)^2/n\}}$$

Prove that T^2 converges in distribution to χ_1^2 as $n_1, n_2, \ldots, n_k \to \infty$ when $\beta = 0$.

6. The data given in Table 4 (p. 27) are the frequencies with which each pair of decimal digits occurs in two parallel sequences of 10,000 digits each. Test the hypothesis that the probability for each digital pair is 0·01.

7. $N_1, N_2, \ldots, N_k$ have a multinomial distribution with total frequency n and probabilities $\pi_1, \pi_2, \ldots, \pi_k$ respectively. Show that the statistic $2\Sigma N_i \log(N_i/n\pi_i)$ converges in distribution to χ_{k-1}^2 as $n \to \infty$.

Now suppose that the probabilities are $p_1(\theta), p_2(\theta), \ldots, p_k(\theta)$ respectively, and let $\hat{\theta}$ be the maximum likelihood estimator of θ. Show that $2\Sigma N_i \log\{N_i/np_i(\hat{\theta})\}$ converges in distribution to χ_{k-2}^2 as $n \to \infty$.

8. The number of yeast cells in a unit area has been assumed in §7 to follow a Poisson distribution with parameter λ. Show that the method of estimating λ by the average number of cells per unit area is in accordance with the principle of maximum likelihood. Test whether a Poisson distribution satisfactorily describes the data by two methods, namely (*a*) using the index of dispersion, and (*b*) using the statistic X^2. Which of procedures (*a*) and (*b*) is preferable for testing the assumption of Poisson probabilities?

9. The arguments of §56 can be generalized for the analysis of problems where the probabilities of the multinomial distribution each depend on two or more parameters. When there are two parameters θ and ψ, denote the probabilities by $p_i(\theta, \psi)$, where $i = 1, 2, \ldots, k$. We then find that

$$X^2 = \sum_{i=1}^{k} \{N_i - np_i(\hat{\theta}, \hat{\psi})\}^2/np_i(\hat{\theta}, \hat{\psi})$$

converges in distribution to χ_{k-3}^2 as $n \to \infty$, where the estimators $\hat{\theta}$ and $\hat{\psi}$ satisfy the equations

$$\sum_{i=1}^{k} N_i \frac{\partial \log p_i(\hat{\theta}, \hat{\psi})}{\partial\theta} = 0 \quad \text{and} \quad \sum_{i=1}^{k} N_i \frac{\partial \log p_i(\hat{\theta}, \hat{\psi})}{\partial\psi} = 0.$$

Apply these results to the problem in exercise 3, and show that

$$X^2 = n(N_{11}N_{22}-N_{12}N_{21})^2/N_{10}N_{20}N_{01}N_{02}.$$

By using the limit theory of the hypergeometric distribution, prove that the test procedure derived in exercise 3 is asymptotically equivalent to the procedure of referring X^2 to the distribution of χ_1^2.

10. The conditions are the same as in exercise 3. Define

$$U = \frac{n^{\frac{1}{2}}(\log N_{11}-\log N_{12}-\log N_{21}+\log N_{22})}{N_{11}^{-1}+N_{12}^{-1}+N_{21}^{-1}+N_{22}^{-1}}.$$

Prove that U converges in distribution to $\mathcal{N}(0, 1)$ as $n \to \infty$ when the criteria of classification are independent. Show also that U^2 differs by $0(n^{-\frac{1}{2}})$ from the variate X^2 obtained in the previous exercise, with a probability which tends to 1 as $n \to \infty$.

BIBLIOGRAPHY

The following books are not mentioned in the text. They deal more comprehensively with the same topics or describe important aspects of statistics which have been excluded for reasons of space. Each is strongly recommended and the contents are summarized below.

COX, D. R. and MILLER, H. D. 1965. *The theory of stochastic processes*. Methuen, London.
This book is concerned with systems which develop in time according to the laws of probability.

CRAMÉR, H. 1946. *Mathematical methods of statistics*. Princeton University Press, New Jersey.
Many basic theorems of probability and statistics are here proved with complete mathematical rigour.

DAVIES, O. L. (Ed.) 1956. *The design and analysis of industrial experiments*. 2nd edn. Oliver and Boyd, Edinburgh.
The several authors give many illustrations of the practice of experimental design.

DE GROOT. M. H. 1970. *Optimal statistical decisions*. McGraw Hill, New York.
The applications of decision theory refer to statistical problems where the concepts of action and utility are relevant.

FISZ, M. 1963. *Probability theory and mathematical statistics*, 3rd ed. Wiley, New York.
This book has the same general approach as the present one but the information given is more detailed.

KENDALL, M. G. and STUART, A. 1958 on. *The advanced theory of statistics*. Vol. 1: *Distribution theory*. Vol. 2: *Inference and relationship*. Vol. 3: *Design and analysis, and time-series*. Griffin, London.
The three volumes of this work survey theoretical statistics with a wealth of illustration and informed discussion.

LINDLEY, D. V. 1965. *Introduction to probability and statistics from a Bayesian viewpoint*. Part 1: *Probability*. Part 2: *Inference*. Cambridge University Press, London.
The two parts give a description of Bayesian analysis, which constitutes an alternative approach to the questions discussed in chapters 7, 8, 9 and 11.

MORAN, P. A. P. 1968. *An introduction to probability theory*. Clarendon Press, Oxford.
The book provides the main facts of probability theory with applications in natural science.

SCHEFFÉ, H. 1959. *The analysis of variance*. Wiley, New York.
This book gives a detailed account of a technique which is widely used.

Statistical Tables

FISHER, R. A. and YATES, F. 1963. *Statistical tables for biological, agricultural and medical research.* 6th ed. Oliver and Boyd, Edinburgh.

GREENWOOD, J. A. and HARTLEY, H. O. 1962. *Guide to tables in mathematical statistics.* Princeton University Press, New Jersey.

PEARSON, E. S. and HARTLEY, H. O. 1966. *Biometrika tables for statisticians*, Vol. 1, 3rd ed. Cambridge University Press, London.

Sources of Data and Author Index

BERKSON, J. 1966. Examination of randomness of α-particle emissions. Pp. 37-54 in *Festschrift for J. Neyman: Research Papers in Statistics.* Wiley, London. [49]

COCHRAN, W. G. 1954. Some methods for strengthening the common χ^2 tests. *Biometrics*, **10**, 417-451. [179]

DRINKWATER, R. W. and HASTINGS, N. A. J. 1967. An economic replacement model. *Opl. Res. Q.*, **18**, 121-138. [30]

EDWARDS, A. W. F. and FRACCARO, M. 1960. Distribution and sequences of sexes in a selected sample of Swedish families. *Ann. hum. Genet.*, **24**, 245-252. [19, 29]

HALL, K. 1948. Further notes on the action of oestrone and relaxin on the pelvis of the spayed mouse, including a single-dose test of potency of relaxin. *J. Endocrinol.*, **5**, 314-321. [140]

LAPATSANIS, P. D. and IRVING, I. M. 1963. A study of specific *E. coli* infections occurring in a unit for surgical neonates. *Acta Paediat., Stockh.*, **52**, 436-442. [187]

LEE, J. A. H. 1963. Seasonal variation in leukaemia incidence. *Br. med. J.*, (II), 623. [182]

MALARIA RESEARCH UNIT 1946. Radiographic studies of the effects of mepacrine on the gastro-intestinal tract of the rat. *Quart. J. exp. Physiol.*, **33**, 311-321. [157]

MARSDEN, E. and BARRATT, T. 1911. The probability distribution of the time intervals of α particles with application to the number of α particles emitted by uranium. *Proc. phys. Soc.*, **23**, 367-373. [37]

STRUTT, J. W., BARON RAYLEIGH 1903. *Scientific Papers.* **4**, 1892-1901. Cambridge University Press. [134]

'STUDENT' 1907. On the error of counting with a haemacytometer. *Biometrika*, **5**, 351-360. [21]

'STUDENT'. 1908. The probable error of a mean. *Biometrika*, **6**, 1-25. [132]

THOMSON, W. E. 1959. ERNIE—a mathematical and statistical analysis. *J. Roy. statist. Soc. A*, **122**, 301-333. [27]

VISTELIUS, A. B. 1960. The skew frequency distributions and the fundamental law of the geochemical processes. *J. Geol.*, **68**, 1-22. [41]

WALLACE, H. R. 1954. Notes on the biology of *Arhopalus ferus* Mulsant (Coleoptera: Cerambycidae). *Proc. Roy. ent. Soc. A*, **29**, 99-113. [145, 156]

SUBJECT INDEX